VISUALIZING

EARTH
SCIENCE

VISUALIZING
EARTH
SCIENCE

Zeeya Merali, PhD

Brian J. Skinner, PhD
Yale University

with contributions by
Alan Strahler, PhD
Boston University

WILEY

In collaboration with
THE NATIONAL GEOGRAPHIC SOCIETY

CREDITS

VP AND PUBLISHER Jay O'Callaghan

MANAGING DIRECTOR Helen McInnis

EXECUTIVE EDITOR Ryan Flahive

DIRECTOR OF DEVELOPMENT Barbara Heaney

MANAGER, PRODUCT DEVELOPMENT Nancy Perry

DEVELOPMENT EDITOR Carolyn Smith

PROJECT EDITOR Joan Kalkut

ASSISTANT EDITOR Courtney Nelson

EDITORIAL ASSISTANTS Erin Grattan, Sean Boda

EXECUTIVE MARKETING MANAGER Jeffrey Rucker

MARKETING MANAGER Danielle Torio

PRODUCTION MANAGER Micheline Frederick

MEDIA EDITORS Lynn Pearlman, Bridget O'Lavin

CREATIVE DIRECTOR Harry Nolan

COVER DESIGNER Harry Nolan

INTERIOR DESIGN Vertigo Design

SENIOR PHOTO EDITOR Elle Wagner

PHOTO RESEARCHER, NATIONAL GEOGRAPHIC Stacy Gold

SENIOR ILLUSTRATION EDITOR Sandra Rigby

PRODUCTION SERVICES Camelot Editorial Services, LLC

Cover credits: Main photo: Stockbyte/SuperStock, Inc.; Thumbnails (from left to right): NASA/JPL/CALTECH/OLIVER KRAUSE/NG Image Collection; Creatas/MediaBakery; Photodisc/SuperStock, Inc.; John Eastcott and Yva Momatiuk/NG Image Collection; Photographers Choice RF/SuperStock, Inc.

This book was set in New Baskerville by Preparé, Inc., and printed and bound by Quebecor World. The cover was printed by Phoenix Color.

To order books or for customer service, please call 1-800-CALL WILEY (225-5945).

ISBN-13: 978-0471-74705-5
BRV ISBN: 978-0470-41847-5

Printed in the United States of America

10 9 8 7 6

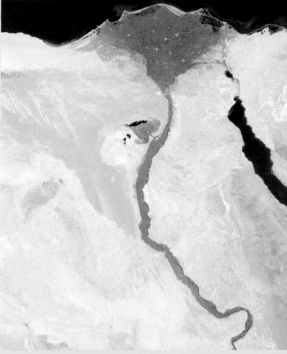

The goal of *Visualizing Earth Science* is *to bring the story of the way Earth works to the majority of students,* not just to those few who choose a career in the sciences. The groundbreaking Visualizing books by John Wiley and Sons in collaboration with the National Geographic Society and its vast image resource, provide an opportunity to tell Earth's story in a fresh new way, while maintaining the rigor and currency needed in science. Students will learn that features we see and experience in the environment around us result from the interactions of many separate phenomena and processes which extend from Earth's core to the outer reaches of the solar system.

This book is intended to serve as an introductory text primarily for undergraduate students who are not majoring in a scientific discipline. The accessible format of *Visualizing Earth Science,* which has been constructed under the assumption that students have little prior knowledge of Earth science, allows students to easily make the transition from an introduction to a topic to its more complex aspects. With its highly visual presentation, which mirrors the very nature of Earth itself, this book is appropriate for use in one-semester courses in Earth science of the kind offered at many institutions of higher learning.

ORGANIZATION

Visualizing Earth Science is structured around two related systems, the Earth system and the solar system. Earth is a system of interacting parts—geosphere, hydrosphere, atmosphere, and biosphere—and as modern science has made it possible to study the interactions within and between the parts in real time, the new field of Earth system science has emerged. But Earth is just one body among many in the still larger solar system, and from that larger system Earth receives the energy that determines weather and climate and powers the biosphere.

Chapter 1 introduces Earth system science, the solar system, the scientific method of investigation, and the reliance of the human population on natural resources, some of which are either locally or globally limited. Chapters 2, 3, and 4 deal with the materials of which the solid Earth is composed—minerals, rocks, and soils. Chapter 5 addresses the hydrosphere, and Chapter 6 regions of climatic extremes—glaciers, ice sheets, and deserts.

The continual restructuring of Earth as a result of forces driven by the internal heat energy is covered in Chapters 7 though 9. Chapter 7 discusses plate tectonics and the ever-changing landscape; Chapter 8 deals with

earthquakes and Earth's internal structure; and Chapter 9 discusses magma, volcanism, and other igneous manifestations.

Earth science is not just Earth system science. Earth science is also the past history of Earth, and Chapters 10 and 11 address this. Chapter 10 covers the determination of relative time and numerical time, and Chapter 11 presents a brief history of life on Earth, from the most ancient evidence to the present.

We walk on the solid Earth but we live in a fluid, the atmosphere, and are surrounded by oceans of seawater. Chapters 12 through 15 are devoted to the fluid regimes. Chapters 12 and 13 are devoted to the oceans and to the shoreline regime, where the ocean meets the land. Chapters 14 and 15 introduce the atmosphere, its structure, circulation, and weather systems.

The final chapters of the book discuss climate and the solar system. Climates past and present, discussed in Chapter 16, attract much attention today as we face the problem of possible climate changes induced by human activities. Space exploration continues to provide a flood of data and new understandings about the solar system, including hypotheses of its origin, its structure today, lessons for Earth from other planets, and the recent discovery that planetary systems exist around other nearby stars. These and other topics are discussed in Chapter 17.

Visualizing Earth Science is, as mentioned earlier, intended as a textbook for an introductory college-level course in Earth science. We do not expect that most of the students who read the book will go on to become Earth scientists, but we hope that all will come to have a better understanding of, and appreciation for, their home planet, its history, and its place in the solar system. For those students who do wish to take further courses in the field—and we hope there are many—we have provided a solid, sufficient, and challenging background to do so with confidence.

NATIONAL GEOGRAPHIC SOCIETY

Visualizing Earth Science offers an array of remarkable photographs, maps, media, and film from the National Geographic Society collections. Students using the book benefit from the long history and rich, fascinating resources of National Geographic.

Fact-Checking: The National Geographic Society has also performed an invaluable service in fact-checking *Visualizing Earth Science*. They have verified every fact in the book with two outside sources, to ensure that the text is accurate and up-to-date.

MEDIA AND SUPPLEMENTS

Visualizing Earth Science is accompanied by a rich array of media and supplements that incorporate the visuals from the textbook extensively to form a pedagogically cohesive package. For example, a Process Diagram from the book appears in the Instructor's Manual with suggestions on using it as a PowerPoint in the classroom; it may be the subject of a short video or an online animation; and it may also appear with questions in the Test Bank, as part of the chapter review, homework assignment, assessment questions, and other online features.

INSTRUCTOR SUPPLEMENTS

WileyPLUS offers a powerful online tool that provides instructors and students with an integrated suite of teaching and learning resources in one easy-to-use Web site. These resources include:

VIDEOS

A collection of videos, a number from the award-winning National Geographic Film Collection, have been selected by Robert Altamura of Florida Community College at Jacksonville to accompany and enrich the text. Each chapter includes at least one video clip, available online as digitized streaming video that illustrates and expands on a concept or topic to aid student understanding. Accompanying each of the videos are contextualized commentary and questions that can further develop student understanding. The videos are available in **WileyPLUS**.

POWERPOINT PRESENTATIONS AND IMAGE GALLERY

A complete set of highly visual PowerPoint presentations by Burair Kothari of Indiana University is available online to enhance classroom presentations. Tailored to the text's topical coverage and learning objectives, these presentations are designed to convey key text concepts, illustrated by embedded text art.

Image Gallery   All photographs, figures, maps, and other visuals from the text are online and can be used as you wish in the classroom. These online electronic files allow you to easily incorporate them into your PowerPoint presentations as you choose, or to create your own overhead transparencies and handouts.

TEST BANK (AVAILABLE IN WILEY*PLUS* AND ELECTRONIC FORMAT)

The visuals from the textbook are also included in the Test Bank by Arthur Lee, Roane State Community College. The Test Bank contains approximately 1700 test items, at least 25 percent of which incorporate visuals from the book. The test items include multiple-choice and essay questions that test a variety of comprehension levels. The Test Bank is available in two formats: online in MS Word files and as a computerized test bank on a multiplatform CD-ROM. The easy-to-use test-generation program fully supports graphics, printed tests, student answer sheets, and answer keys. The software's advanced features allow you to create an exam to your exact specifications.

INSTRUCTOR'S MANUAL (AVAILABLE IN ELECTRONIC FORMAT)

The Instructor's Manual begins with a special introduction on *Using Visuals in the Classroom*, prepared by Matthew Leavitt of Arizona State University, in which he provides guidelines and suggestions on how to use the visuals in teaching the course. For each chapter, materials by Paul Cutlip of St. Petersburg College include suggestions and directions for using Web-based learning modules in the classroom and for homework assignments, as well as creative ideas for in-class activities.

WEB-BASED LEARNING MODULES

A robust suite of multimedia learning resources have been designed for *Visualizing Earth Science*, focusing on and using the visuals from the book. Available in **WileyPLUS**, the content is organized into *tutorial animations*. These animations visually support the learning of a difficult concept, process, or theory, many of them built around a specific feature such as a Process Diagram, Visualizing feature, or key visual in the chapter. The animations go beyond the content and visuals presented in the book, providing additional visual examples and descriptive narration.

VIEW THIS IN ACTION
in your WileyPLUS course

A number of pedagogical features using visuals have been developed specifically for *Visualizing Earth Science*. Presenting the highly varied and often technical concepts woven throughout Earth science raises challenges for reader and instructor alike. This **Illustrated Book Tour** provides a guide to the diverse features that contribute to *Visualizing Earth Science's* pedagogical plan.

CHAPTER INTRODUCTIONS illustrate certain concepts in the chapter with concise stories about some of the world's most remarkable places and events of unusual interest. These narratives are featured alongside striking photographs. The chapter openers also include illustrated **CHAPTER OUTLINES** that use thumbnails of illustrations from the chapter to refer visually to the content.

GLOBAL LOCATOR MAPS, prepared specifically for this book by the National Geographic Society cartographers, accompany some photos. These locator maps help students visualize where the area depicted in the photo is situated on Earth.

VISUALIZING features are specially designed, multipart visual spreads that focus on a key concept or topic in the chapter, exploring it in detail or in broader context using a combination of photos and figures.

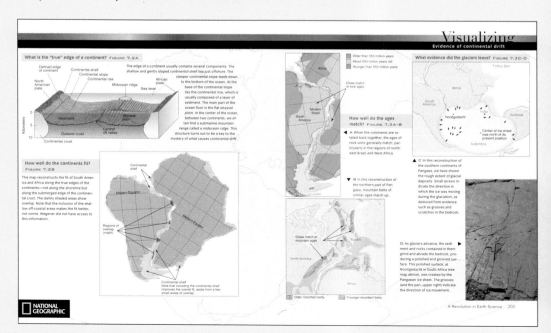

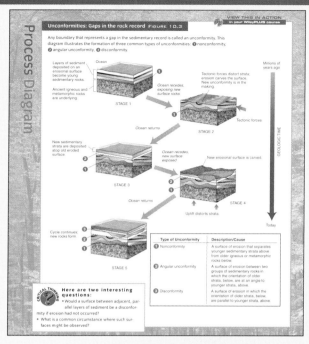

PROCESS DIAGRAMS present a series of figures or a combination of figures and photos that describe and depict a complex process, helping students to observe, follow, and understand the process.

HERE'S AN INTERESTING QUESTION asks students challenging questions related to the topics discussed in What an Earth Scientist Sees and in Process Diagrams.

WHAT AN EARTH SCIENTIST SEES are features that highlight a concept or phenomenon, using photos and figures that would stand out to a professional in the field, and helping students to develop observational skills.

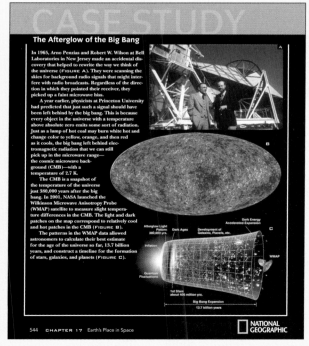

CASE STUDIES are illustrated features that offer a wide variety of in-depth examinations that address important issues in Earth science.

AMAZING PLACES is a feature that takes students to a unique location in the world that provides a vivid illustration of a theme in the chapter. Students could easily visit most of the Amazing Places someday and so continue their Earth science education after they finish this book.

ix

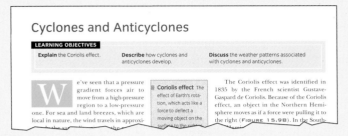

Cyclones and Anticyclones

LEARNING OBJECTIVES

Explain the Coriolis effect.

Describe how cyclones and anticyclones develop.

Discuss the weather patterns associated with cyclones and anticyclones.

We've seen that a pressure gradient forces air to move from a high-pressure region to a low-pressure one. For sea and land breezes, which are local in nature, the wind travels in approxi-

Coriolis effect The effect of Earth's rotation, which acts like a force to deflect a moving object on the surface to the right

The Coriolis effect was identified in 1835 by the French scientist Gustave-Gaspard de Coriolis. Because of the Coriolis effect, an object in the Northern Hemisphere moves as if a force were pulling it to the right (**FIGURE 15.9B**). In the South-

LEARNING OBJECTIVES at the beginning of each section indicate in behavioral terms that the student must be able to demonstrate mastery of the material in the chapter.

sometimes rapidly, thus creating weather disturbances.

nown as **cyclones**, air spirals . inward-spiraling motion is -pressure centers, known as **nticyclones**, air spirals down-ard and outward. This out-ard-spiraling motion is called *vergence.*

Low-pressure centers (cy-ones) are often associated th cloudy or rainy weather, hereas high-pressure centers nticyclones) are often asso-

CONCEPT CHECK STOP

What is the cause of the Coriolis effect?

What are cyclones and anticyclones? **How** do they develop?

How does the Coriolis force deflect the paths of moving objects?

What weather patterns are associated with cyclones and anticyclones?

CONCEPT CHECK questions at the end of each section give students the opportunity to test their comprehension of the learning objectives.

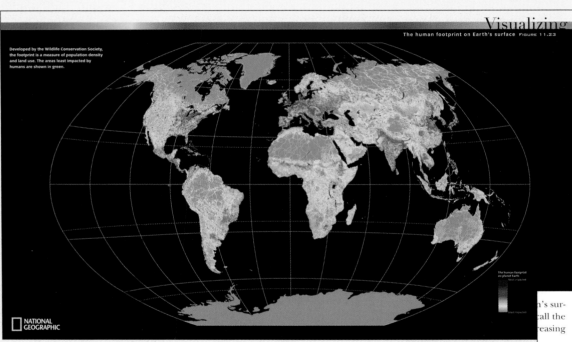

Visualizing

The human footprint on Earth's surface FIGURE 11.23

Developed by the Wildlife Conservation Society, the footprint is a measure of population density and land use. The areas least impacted by humans are shown in green.

The human footprint on planet Earth

NATIONAL GEOGRAPHIC

NATIONAL GEOGRAPHIC SOCIETY MAPS are featured throughout the book.

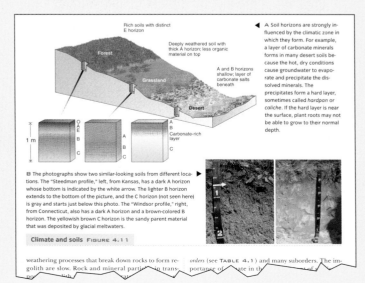

Rich soils with distinct E horizon

Forest

Deeply weathered soil with thick A horizon; less organic material on top

A and B horizons shallow; layer of carbonate salts beneath

Grassland

Desert

A B Carbonate-rich layer C

◀ A Soil horizons are strongly influenced by the climatic zone in which they form. For example, a layer of carbonate minerals forms in many desert soils because the hot, dry conditions cause groundwater to evaporate and precipitate the dissolved minerals. The precipitates form a hard layer, sometimes called *hardpan* or *caliche*. If the hard layer is near the surface, plant roots may not be able to grow to their normal depth.

B The photographs show two similar-looking soils from different locations. The "Steedman profile," left, from Kansas, has a dark A horizon whose bottom is indicated by the white arrow. The lighter B horizon extends to the bottom of the picture, and the C horizon (not seen here) is gray and starts just below this photo. The "Windsor profile," right, from Connecticut, also has a dark A horizon and a brown-colored B horizon. The yellowish brown C horizon is the sandy parent material that was deposited by glacial meltwaters.

Climate and soils FIGURE 4.11

ILLUSTRATIONS AND PHOTOS support concepts covered in the text, elaborate on relevant issues, and add visual detail. Many of the photos originate from National Geographic's rich sources.

lapse rate The rate at which air temperature decreases with increasing altitude.

troposphere The lowest layer of the atmosphere, in which temperature falls steadily with increasing altitude.

MARGIN GLOSSARY TERMS (in green boldface) introduce each chapter's most important terms. The second most important terms appear in **black boldface** and are defined in the text.

weathering processes that break down rocks to form re-golith are slow. Rock and mineral partie in trans-

orders (see **TABLE 4.1**) and many suborders. The im-portance of ate in th

TABLES AND GRAPHS, with data sources cited at the end of the text, summarize and organize important information.

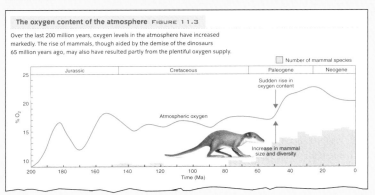

The oxygen content of the atmosphere FIGURE 11.3

Over the last 200 million years, oxygen levels in the atmosphere have increased markedly. The rise of mammals, though aided by the demise of the dinosaurs 65 million years ago, may also have resulted partly from the plentiful oxygen supply.

LF-TESTS at the end of each chapter provide a series of ltiple-choice questions, many of them incorporating visuals m the chapter, that review the major concepts.

SELF-TEST

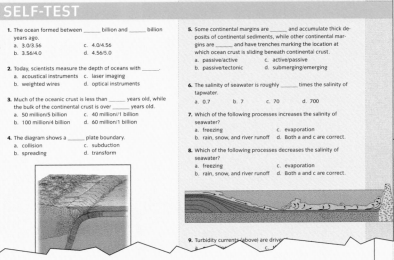

1. The ocean formed between _____ billion and _____ billion years ago.
   a. 3.0/3.56   c. 4.0/4.56
   b. 3.56/4.0   d. 4.56/5.0

2. Today, scientists measure the depth of oceans with _____.
   a. acoustical instruments   c. laser imaging
   b. weighted wires   d. optical instruments

3. Much of the oceanic crust is less than _____ years old, while the bulk of the continental crust is over _____ years old.
   a. 50 million/5 billion   c. 40 million/1 billion
   b. 100 million/4 billion d. 60 million/1 billion

4. The diagram shows a _____ plate boundary.
   a. collision   c. subduction
   b. spreading   d. transform

5. Some continental margins are _____ and accumulate thick deposits of continental sediments, while other continental margins are _____ and have trenches marking the location at which ocean crust is sliding beneath continental crust.
   a. passive/active   c. active/passive
   b. passive/tectonic d. submerging/emerging

6. The salinity of seawater is roughly _____ times the salinity of tapwater.
   a. 0.7   b. 7   c. 70   d. 700

7. Which of the following processes increases the salinity of seawater?
   a. freezing   c. evaporation
   b. rain, snow, and river runoff   d. Both a and c are correct.

8. Which of the following processes decreases the salinity of seawater?
   a. freezing   c. evaporation
   b. rain, snow, and river runoff   d. Both a and c are correct.

9. Turbidity currents (above) are driven _____

CRITICAL AND CREATIVE THINKING QUESTIONS

1. When the height between high and low tide is large, as in the Bay of Fundy, engineers sometimes consider using the tide as a way to generate electricity. What might some of the negative consequences be from the installation of tidal power plants?

2. If you live or go to school near a coastline, identify the kind of coastline that is near. How will the rise of a meter in sea level affect your area?

3. Scientists who study climate predict that as the climate gets warmer, severe storms and particularly hurricanes, will become stronger and more frequent. Which regions of North America are most vulnerable and likely to be affected by the changes?

4. When building a house, it is important to assess potential hazards. What advice would you offer to someone planning to build a shoreline home in the following places: (a) southern Georgia; (b) Oregon, south of the Columbia River; (c) the Yucatan Peninsula, Mexico; (d) the south side of Nantucket Island, Massachusetts.

5. Tsunamis have been recorded in the Atlantic Ocean, but they are much less common than in the Pacific and Indian Oceans. Why is this so?

CRITICAL AND CREATIVE THINKING QUESTIONS encourage critical thinking and highlight each chapter's important concepts and applications.

SUMMARY

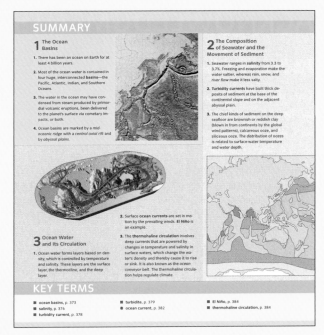

1 The Ocean Basins

1. There has been an ocean on Earth for at least 4 billion years.

2. Most of the ocean water is contained in four huge, interconnected basins—the Pacific, Atlantic, Indian, and Southern Oceans.

3. The water in the ocean may have condensed from steam produced by primordial volcanic eruptions, been delivered to the planet's surface via cometary impacts, or both.

4. Ocean basins are marked by a mid-oceanic ridge with a central axial rift and by abyssal plains.

3 Ocean Water and Its Circulation

1. Ocean water forms layers based on density, which is controlled by temperature and salinity. These layers are the surface layer, the thermocline, and the deep layer.

2 The Composition of Seawater and the Movement of Sediment

1. Seawater ranges in salinity from 3.3 to 3.7%. Freezing and evaporation make the water saltier, whereas rain, snow, and river flow make it less salty.

2. Turbidity currents have built thick deposits of sediment at the base of the continental slope and on the adjacent abyssal plain.

3. The chief kinds of sediment on the deep seafloor are brownish or reddish clay (blown in from continents by the global wind patterns), calcareous ooze, and siliceous ooze. The distribution of oozes is related to surface-water temperature and water depth.

2. Surface ocean currents are set in motion by the prevailing winds. El Niño is an example.

3. The thermohaline circulation involves deep currents that are powered by changes in temperature and salinity in surface waters, which change the water's density and thereby cause it to rise or sink. It is also known as the ocean conveyor belt. The thermohaline circulation helps regulate climate.

KEY TERMS

■ ocean basins, p. 373
■ salinity, p. 376
■ turbidity current, p. 378
■ turbidite, p. 379
■ ocean current, p. 382
■ El Niño, p. 384
■ thermohaline circulation, p. 384

The **SUMMARY** revisits each learning objective and redefines each margin glossary term, featured in boldface here and included in a list of **KEY TERMS**. Students are thus able to study vocabulary words in the context of related concepts. Each portion of the Summary is illustrated with a relevant photo from its respective chapter section.

What is happening in these pictures ?

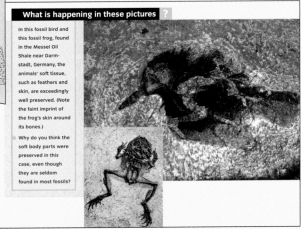

In this fossil bird and this fossil frog, found in the Messel Oil Shale near Darmstadt, Germany, the animals' soft tissue, such as feathers and skin, are exceedingly well preserved. (Note the faint imprint of the frog's skin around its bones.)

■ Why do you think the soft body parts were preserved in this case, even though they are seldom found in most fossils?

WHAT IS HAPPENING IN THIS PICTURE? are end-of-chapter features that present students with a photograph that is relevant to chapter topics but illustrates a situation students are not likely to have encountered previously. The photograph is paired with questions designed to stimulate creative thinking.

ACKNOWLEDGMENTS

PROFESSIONAL FEEDBACK

Throughout the process of writing and developing this text and the visual pedagogy, we benefited from the comments and constructive criticism provided by the instructors and colleagues listed below. We offer our sincere appreciation to these individuals for their helpful reviews:

Robert Altamura
Florida Community College at Jacksonville

Steve Bennett
Western Illinois University

Callan Bentley
Northern Virginia Community College

Natalie Bursztyn
Bakersfield College

Marianne Caldwell
Hillsborough Community College

James Carew
College of Charleston

Claire Coyne
Santa Ana College

Constantin Cranganu
Brooklyn College of the City University of New York

Vincent Devlahovich
California State University, Northridge

Gerald Grams
Clark Atlanta University

Clay Harris
Middle Tennessee State University

George Hazelton
Chowan College

Joann Hochstein
Central Florida Community College

Paul Horton
Indian River Community College

Asaad Istephan
Madonna University

Linda Jones
Southwest Minnesota State University

Burair Kothari
Wilbur Wright College

Kristine Larsen
Central Connecticut State University

Jay Lennartsen
University of North Carolina, Greensboro

Michael Lewis
University of North Carolina, Greensboro

Zhaohui Li
University of Wisconsin, Parkside

Tim Long
Georgia Institute of Technology, Atlanta

Donald Lovejoy
Palm Beach Atlantic University

Steven Maier
Northwest Oklahoma State University

Ravi Nandigan
University of Texas, Brownsville

James F. Nugent
Salve Regina University

Guillermo Rocha
Brooklyn College of the City University of New York

Laura Sanders
Northeastern Illinois University

Steven Schimmrich
SUNY Ulster County Community College

John Tacinelli
Rochester Community and Technical College

Dave Thomas
Washtenaw Community College

Husan Thompson
Loyola Marymount University

Craig Van Boskirk
Florida Community College at Jacksonville

Kim Van Scoy
University of the Ozarks

Jaehyung Yu
Texas A&M University, Kingsville

SPECIAL THANKS

We are extremely grateful to the many members of the editorial and production staff at John Wiley & Sons who guided us through the challenging steps of developing this book. Their tireless enthusiasm, professional assistance, and endless patience smoothed the path as we found our way. We thank in particular Ryan Flahive, who expertly launched and directed our process; Nancy Perry, Manager, Product Development; Helen McInnis, Managing Director, Wiley Visualizing, who oversaw the concept of the book; Carolyn Smith, Developmental Editor, for her careful editing of our book; Micheline Frederick, Production Manager, who stepped in whenever we needed expert advice; Jay O'Callaghan, Vice President and Publisher, who oversaw the entire project; and Jeffrey Rucker, Executive Marketing Manager for

Wiley Visualizing, and Danielle Torio, Marketing Manager for Geosciences, who adeptly represent the Visualizing imprint. We appreciate the expertise of Elle Wagner, Senior Photo Editor, in managing and researching our photo program and of Sandra Rigby, Senior Illustration Editor, in managing the illustration program. We also wish to thank those who worked on the media and ancillary materials: Lynn Pearlman, Senior Media Editor; Bridget O'Lavin; and Erin Grattan.

We wish to thank Barbara Murck, University of Toronto, co-author of *Visualizing Geology*, a book that was the principal source for topics covered in the first half of the book. We are grateful to Stacy Gold, Research Editor and Account Executive at the National Geographic Image Collection, for her valuable expertise in selecting NGS photos. Many other individuals at National Geographic offered their expertise and assistance in developing this book: Richard Easby, Executive Editor, National Geographic School Division; Mimi Dornack, Sales Manager, and Lori Franklin, Assistant Account Executive, National Geographic Image Collection; and Dierdre Bevington-Attardi, Project Manager, and Kevin Allen, Director of Map Services, National Geographic Maps.

ABOUT THE AUTHORS

Zeeya Merali has an undergraduate degree and a Master's in natural sciences from the University of Cambridge, and a PhD in theoretical cosmology from Brown University. She also holds a Master's degree in science communication from Imperial College, London. As a science writer, her work has appeared in *Scientific American* magazine, the journal *Nature*, and *New Scientist* magazine, and as a filmmaker, her work has been broadcast on The History Channel, UK.

Brian Skinner was born and raised in Australia, studied at the University of Adelaide in South Australia, worked in the mining industry in Tasmania, and in 1951 entered the Graduate School of Arts and Sciences, Harvard University, from which he obtained his PhD in 1954. Following a period as a research scientist in the United States Geological Survey in Washington, DC, he joined the faculty at Yale in 1966, where he continues his teaching and research as the Eugene Higgins Professor of Geology and Geophysics. Brian Skinner has been president of the Geochemical Society, the Geological Society of America, and the Society of Economic Geologists. He holds an honorary Doctor of Science from Toronto University and an honorary Doctor of Engineering from the Colorado School of Mines.

Alan Strahler earned his PhD in geography from Johns Hopkins in 1969, and is presently Professor of Geography at Boston University. He has published over 250 articles in the refereed scientific literature, largely on the theory of remote sensing of vegetation, and has also contributed to the fields of plant geography, forest ecology, and quantitative methods. In 1993, he was awarded the Association of American Geographers/Remote Sensing Specialty Group Medal for Outstanding Contributions to Remote Sensing. With Arthur Strahler, he is co-author of 7 textbook titles with 11 revised editions on physical geography and environmental science. He holds the honorary degree DSHC from the Université Catholique de Louvain, Belgium, and is a Fellow of the American Association for the Advancement of Science.

CONTENTS *in Brief*

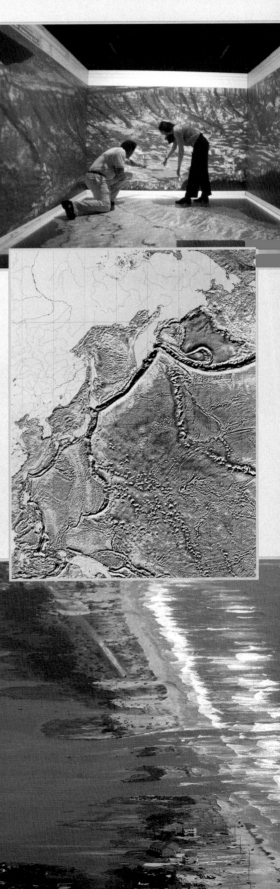

CONTENTS

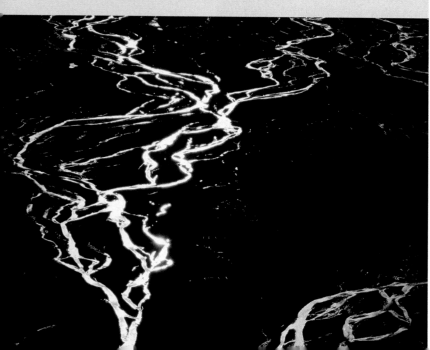

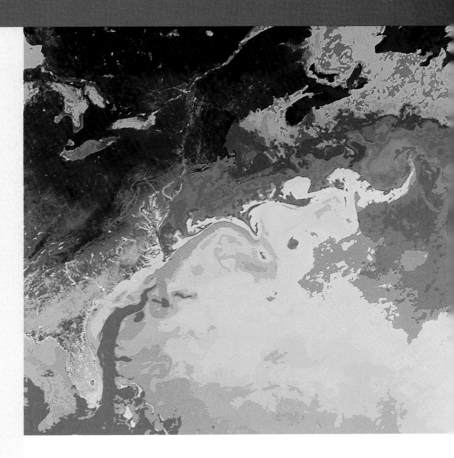

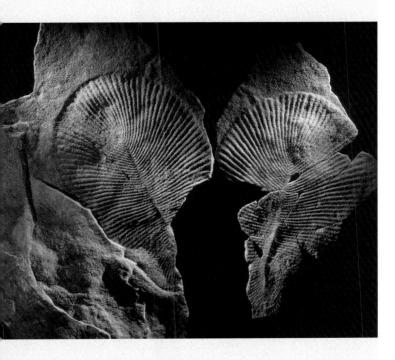

VISUALIZING FEATURES

Multipart visual presentations that focus on a key concept or topic in the chapter.

PROCESS DIAGRAMS

A series or combination of figures and photos that describe and depict a complex process.

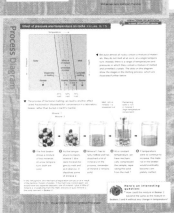

Achieve Positive Learning Outcomes

1 What Is Earth Science?

1. **Earth science** is the study of all aspects of Earth. Earth scientists study the record contained in rocks of all that has happened in the past, the interactions between all parts of the Earth system today, and the probable future changes to the environment in which we live as a consequence of our collective human activities.

2. The **scientific method** is a research strategy that scientists use to study a problem by formulating a hypothesis and then testing it by performing an experiment. The steps include (1) observing and gathering data; (2) formulating a **hypothesis**; (3) testing the hypothesis; (4) formulating a **theory**; and (5) formulating a **law** or **principle**.

3. Earth scientists study Earth today using **Earth system science**. This concept comes from the discovery that Earth is an integrated **system** of interconnected and interdependent parts. Individual systems within the larger Earth system

can be big or small, and can vary greatly in complexity, but regardless of size, each system operates within an identifiable boundary. There are three kinds of systems: *isolated, closed,* and *open;* the properties of the boundary determine the kind of system. In an isolated system, boundaries prevent the system from taking in or releasing any energy or matter. Because there is no perfect boundary against the passage of energy, isolated systems do not exist in the real world. A closed system has a boundary that permits the passage of energy, but not of matter, in and out of the system. The third kind of system, an *open system*, permits the exchange of both matter and energy across its boundary. Most environmental and geologic systems in the natural world are open systems.

4. Earth is considered a closed system, though some small amounts of matter do cross its boundary. The Earth sys-

tem consists of four principal open subsystems, including the **atmosphere**, the envelope of gas that surrounds Earth; the **hydrosphere**, comprising of all the Earth's water; the **biosphere**, all of Earth's living organisms; and the **lithosphere**, Earth's rocky outer layer. Materials and energy are stored for varying lengths of time in each of these systems or reservoirs and can move among them via innumerable pathways and processes. Each of the four Earth subsystems can be further broken down into a vast number of still smaller subsystems, all of which are open.

5. An important component of Earth system science is the monitoring and study of the movement of materials among the subsystems. The system is kept in balance, or reaches a new balance following a change in some part of the system through feedback mechanisms. **Positive feedback mechanisms** work to change the system; **negative feedback mechanisms** work to resist change. Feedbacks are especially important in the **life zone**, the region between 10 km above and 10 km below sea level, where all of life on Earth is located.

2 Earth in Space

1. Earth is one of the eight bodies in the **solar system** recognized as planets. In addition to the Sun and *planets*, the solar system includes a vast number of moons, asteroids, comets, and fragments of rock called *meteoroids*. The four inner planets, or *terrestrial planets*, Mercury, Venus, Earth, and Mars, are similar in many ways. They are all small, rocky, and relatively dense, and they have similar sizes and chemical compositions. The four outer planets, or *Jovian planets*, in contrast, consist of huge gaseous atmospheres with small solid cores, giving them very low densities overall. The Jovian planets are Jupiter, Saturn, Uranus, and Neptune. Pluto, until recently considered to be the ninth planet, is not a Jovian planet because it is icy but not gassy. Instead, it is considered to be a "dwarf" planet and part of the Kuiper Belt, a region of the outer edge of the solar system that contains a large number of small, icy objects.

2. Early in its history, Earth underwent *differentiation* into a dense, metallic **core**, a rocky **mantle**, and a brittle, rocky outer **crust**. Because of the way in which temperature and pressure control the strength of rocks, the outermost 100 km or so of the solid Earth—that is the crust and upper part of the mantle—consists of rocks that are tough and resistant to breakage; this zone is called the **lithosphere**. Beneath the lithosphere, from a depth of about 100 km to a depth of about 350 km, is a zone where rocks are weak and easily deformed, and though not actually molten, so ductile they behave like very thick liquids. This zone is called the **asthenosphere**. The great bulk of the mantle, from a depth of 350 km to the boundary between the mantle and the core, is called the *mesosphere*, and it consists of rocks that are readily deformed but are not as ductile as the asthenosphere.

3. Earth is unique in the solar system in that it possesses an oxygen-rich atmosphere. Earth is also the only planet in the solar system with a hydrosphere in which water exists near the surface in solid, liquid, and gaseous forms, and a biosphere with living organisms. Finally, Earth is the only planet where true soil is formed from *regolith* by interactions among physical, chemical, and biologic processes, and where life as we know it could exist.

4. **Plate tectonics** is the motion and interaction of large segments of the lithosphere. It is because of plate tectonics that Earth has two fundamentally different types of crust: the relatively thin, dense **oceanic crust** of the volcanic rock *basalt* and the thicker, less dense **continental crust** comprised mainly of the igneous rock *granite*.

3 Humans and Earth

1. Study of human history reveals more than one example of societies that have collapsed because of a failure to use limited natural resources wisely. The Easter Islanders in the South Pacific and the Garamantes in North Africa are two examples.

2. **Natural resources**, which include all of the materials we take from the Earth system, can be divided into two families, the **renewable resources** and the **nonrenewable resources**. The renewables include materials such as water in streams and agricultural products, which are continually replaced or can be newly produced each growing season.

Nonrenewables are those resources that cannot be regenerated on human timescales and so are one-crop materials.

3. The study of Earth science is important to human society for many reasons.

Earth materials and processes affect our lives through our dependence on Earth resources: through geologic hazards such as volcanic eruptions, floods, and earthquakes; and through the physical properties of the natural environment.

KEY TERMS

CRITICAL AND CREATIVE THINKING QUESTIONS

1. Do you think there may be life on a planet outside of our solar system? What would the atmosphere of that planet be like? Must it have a hydrosphere? Why or why not?

2. Why is the systems approach so useful in studying both natural and artificial processes? Can you think of examples of artificial (that is, human-built) systems other than those given in the text? Are they open systems or closed systems? (Think about the materials and energy in them.)

3. In this chapter we have suggested that Earth is a close approximation of a natural closed system, and we have hinted at some of the ways that living in a closed system affect each of us. Can you think of some other ways?

4. In what ways do Earth science processes affect your daily life?

5. Formation of clouds and removal of carbon from the atmosphere by growing plants are negative feedback mechanisms tending to cool Earth's atmosphere. Can you think of other negative feedback mechanisms that might involve the ocean or some other part of the Earth system?

6. How many things on which you rely for your daily activities require the use of nonrenewable resources? All nonrenewable resources in a closed system are limited; which ones do you think might have limits that will affect the long-term activities of the human population?

What is happening in this picture ?

This rock, photographed in Saudi Arabia's Rhub al Khali (Empty Quarter), was discovered in 1965. It is believed to be the largest fragment of a meteorite that fell to Earth sometime before 1863 (when the first piece was discovered).

- How do you think these scientists can tell it is a meteorite?

- Why did it break up into pieces?

- Why is the desert a good place to look for meteorites?
  (Hint: Think about what would have happened to this rock if it had fallen in a jungle or a mountain range.)

SELF-TEST

1. Earth science is the scientific study of _____.
- a. soils
- b. rocks and minerals
- c. all aspects of Earth
- d. all the terrestrial planets
- e. mines and oil fields

2. The scientific method is a way of investigating natural phenomena by _____.
- a. challenging entrenched beliefs
- b. using mathematics to confuse students
- c. defending personal ideas
- d. making observations and repeatedly testing conclusions
- e. taking photographs from space

3. On this illustration, label each of the following systems:

    isolated system
    closed system
    open system

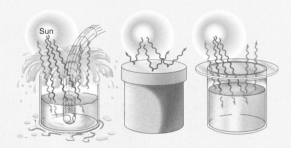

4. The island depicted in the figure acts as a(n) _____.
- a. intermittent system
- b. closed system
- c. solar system
- d. open system
- e. isolated system

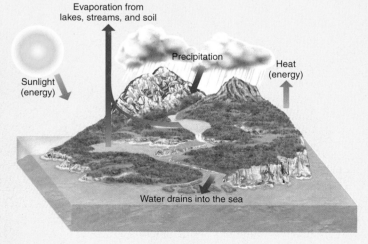

5. On the time scale of a human lifetime, Earth acts as a(n) _____.
- a. intermittent system
- b. closed system
- c. solar system
- d. open system
- e. isolated system

6. The _____ is a subset of the Earth system that comprises all of its bodies of water and ice, both on the surface and underground.
- a. atmosphere
- b. hydrosphere
- c. lithosphere
- d. ionosphere
- e. biosphere

7. Earth's climate is balanced by _____.
- a. the growth of forests acting as a negative feedback
- b. the absence of any negative feedbacks
- c. clouds of carbon dioxide acting as a negative feedback
- d. an interplay between positive and negative feedbacks
- e. the absence of any positive feedbacks

8. On this illustration label the following objects:

Venus, Earth, Mars, Jupiter, the terrestrial planets, Neptune, asteroids, Pluto, Saturn, Kuiper Belt objects

9. The photograph is of a basalt flow on the island of Hawaii. Which one of the following statements is true?

a. Basaltic lava flows have only occurred on Earth.

b. These types of flows have been common only on Earth and in the early history of the Moon.

c. Basalt is the most common volcanic rock known in our solar system.

d. Basaltic lava flows would have been common in the early history of Earth, but in modern times they have largely ceased.

e. None of the above statements is true.

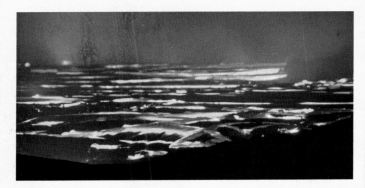

10. Earth, Mars, and Venus all have _____.

a. an oxygen- and nitrogen-rich atmosphere

b. a core, mantle, and crust

c. a hydrosphere with liquid water

d. a biosphere

e. All of the above statements are true.

11. On this illustration, label Earth's internal structure using the following terms:

mantle	lithosphere	oceanic crust
asthenosphere	outer core	inner core
continental crust	mesosphere	

12. The asthenosphere is a layer whose distinctiveness from the rest of the mantle is based on its _____.

a. difference in composition

b. high strength

c. low strength

d. relatively low temperature

e. increased brittleness

13. All known forms of life live in a restricted zone where the hydrosphere, atmosphere, and lithosphere interact to provide the right balance of physical conditions and nutrients; this life zone is _____.

a. between the Arctic circle and the Antarctic circle

b. from the shoreline to the top of the atmosphere

c. between 20 km below and 20 km above sea level

d. between 10 km below and 10 km above sea level

e. between 5 km below and 5 km above sea level

14. Which one of the following statements is incorrect?

a. The depletion of natural resources apparently led to the collapse of some ancient civilizations.

b. Nonrenewable resources are never replenished.

c. Renewable resources must be managed so they are not used at a rate that is greater than the rate of renewal or replenishment of the resource.

d. Groundwater is, in principle, a renewable resource, but once depleted it may take a very long time to be replenished.

e. Fossil fuels are nonrenewable resources, but metals are renewable resources.

15. The study of Earth science is important because _____.

a. it helps us understand the processes that govern the Earth system

b. it helps us assess the potential limitations of the supplies of natural resources on which civilization depends

c. it helps us understand and mitigate the potential threats of natural hazards, such as floods, landslides, earthquakes, volcanic eruptions, and even meteorite impacts

d. it makes us more aware of the uniqueness of this planet that we share with all other life-forms

e. All of the above statements are true.

Minerals: Earth's Building Blocks

This diamond comes from Point Lake, Northwest Territories, Canada, where geologists Charles Fipke and Stewart Blusson discovered a rich diamond deposit in 1991. Fipke and Blusson hypothesized that diamonds found in Wisconsin had been carried there by glaciers during the last Ice Age. They were proved right. Since 1998, when the first mine opened, Canada has become the world's third largest exporter of diamonds. Every diamond produced in Canada is engraved with a tiny polar bear that attests to its origin.

Diamond is a remarkable mineral. It is pure elemental carbon, the same chemical found in graphite and charcoal. But, unlike graphite and charcoal, natural diamond forms only at extraordinarily high pressures and temperatures, deep under Earth's surface or at the center of meteorite impacts. It is the hardest mineral known and an excellent conductor of heat. Its properties have made it a valuable material for industrial purposes. If production costs can be brought down, artificial diamonds might one day replace silicon in our computers, which could then run at much higher temperatures.

In this chapter you will learn about minerals and why minerals of identical composition, such as diamond and graphite, have different properties. Some minerals are beautiful, some are economically important (but not beautiful), and some are vital as nutrients. A few minerals, such as certain varieties of asbestos, are potentially hazardous to human health. Earth scientists need to study minerals and their properties, so that we can learn to balance the positive and negative effects on our lives.

Point Lake

Global Locator

NATIONAL GEOGRAPHIC

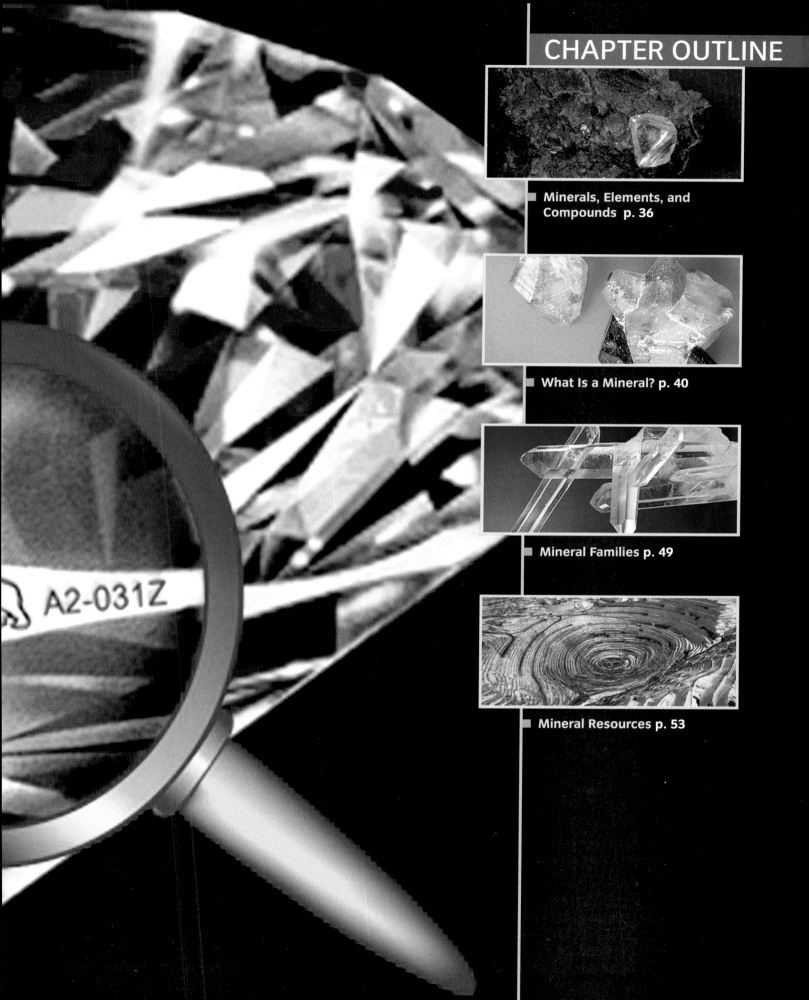

A2-031Z

Minerals, Elements, and Compounds

LEARNING OBJECTIVES

State the four requirements for a solid material to be classified as a mineral.

Define element, atom, compound, molecule, and ion.

Explain the difference between an atom and a molecule.

Describe the internal structure of an atom.

Identify four kinds of chemical bonding.

Explain how the kinds of bond in a material affect the properties of a mineral.

The word *mineral* has a specific connotation in Earth science. To be classified as a mineral, a substance must meet certain criteria. A **mineral** must

- Be a naturally occurring solid.
- Be formed by inorganic processes.
- Have a characteristic crystal structure.
- Have a specific chemical composition.

Each criterion is essential, and because minerals are solids, we will first discuss how solids form. Do not confuse a mineral with a rock. A **rock** is a solid aggregate of minerals, and in the next chapter we discuss how minerals combine to form rocks. In this chapter, we concentrate on minerals, and to do so we start with the fundamental particles that are present in all minerals—atoms.

> **mineral** A naturally formed, solid, inorganic substance with a characteristic crystal structure and a specific chemical composition.

> **rock** A naturally formed, coherent aggregate of minerals and possibly other nonmineral matter.

ELEMENTS, ATOMS, AND IONS

Chemical **elements** are the most fundamental substances into which matter can be separated and analyzed by ordinary chemical methods. All matter on Earth, including the page you are reading and the eyes you are reading it with, consists of one or more chemical elements. All of the chemical reactions that make life on Earth possible depend on the ways chemical elements interact. Ninety-two naturally occurring elements are known, and a number more have been synthesized by atomic scientists. Each element is identified by abbreviated symbols, such as H for hydrogen and Si for silicon. Some of the symbols come from other languages, such as Fe for iron, from the Latin *ferrum,* and Na for sodium, from the Latin *natrium.* Others are named in honor of famous scientists, such as element 99, Es, einsteinium. The periodic table of the elements is shown in Appendix A.

> **element** The most fundamental substance into which matter can be separated by chemical means.

> **atom** The smallest individual particle that retains the distinctive chemical properties of an element.

Even the tiniest grain of dust is made up of innumerable particles, called **atoms**, which are much too small to see (**FIGURE 2.1**). They are so tiny, about one-billionth of a millimeter, that they cannot even be seen at all with an optical microscope. Special microscopes that do not use light have succeeded in imaging atoms, but it would be more accurate to say that they "feel" the atoms rather than see them. It may seem strange that the properties of things as tiny as atoms should determine the properties of Earth, but they do.

Chemical reactions take place between atoms, and it is those reactions that produce the minerals, liquids, and gases that Earth scientists study. Atoms themselves are composed of even smaller particles, which have no independent chemical properties. The *nucleus* (plural *nuclei*) of an atom contains *protons,* with positive electric charges, and *neutrons,* which are electrically neutral. The number of protons in an atom—its *atomic number*—determines its chemical characteristics. Atomic numbers range from 1 for the lightest element, hydrogen, up to 92 for uranium, the heaviest naturally occurring element. Every element from atomic number 1 to 92

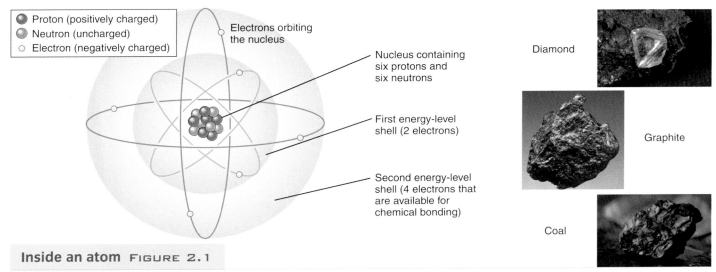

A Single Atom of Carbon-12 (Schematic Diagram)

- ● Proton (positively charged)
- ● Neutron (uncharged)
- ○ Electron (negatively charged)

Electrons orbiting the nucleus

Nucleus containing six protons and six neutrons

First energy-level shell (2 electrons)

Second energy-level shell (4 electrons that are available for chemical bonding)

Three things that contain carbon

Diamond

Graphite

Coal

Inside an atom FIGURE 2.1

As shown in the diagram of carbon-12 (left), six electrons orbit the nucleus in two complex paths called *orbitals,* rendered here (unrealistically) as circles. The orbitals arrange themselves in energy-level shells, which are more stable when completely filled—the first energy-level shell is filled when two electrons are present, the second shell can have eight electrons. There are two carbon isotopes, carbon-12 (the major component) and carbon-13, present in diamond, graphite, and coal. All living beings also contain carbon-12 and carbon-13, but in addition they contain trace amounts of carbon-14, a radioactive isotope.

has either been synthesized in the laboratory or found in nature, so there are no new elements to be discovered in that range. However, scientists are working on synthesizing heavier elements and have reached element 118 (ununoctium).

The number of protons plus the number of neutrons in the nucleus of an atom is the *mass number.* Atoms of a given element always have the same atomic number, but they can have different mass numbers. For

| **isotopes** Atoms with the same atomic number and different mass numbers. |

example, there are three naturally occurring **isotopes** of carbon: carbon-12, carbon-13, and carbon-14. Each of the isotopes of carbon has 6 protons and thus an atomic number of 6. However, the three isotopes contain different numbers of neutrons: 6, 7, and 8 per atom, respectively (thus, different mass numbers: 12, 13, and 14).

The third component of an atom is called an *electron.* Electron interactions help determine the makeup of ions and compounds. Electrons orbit the nucleus, as shown schematically in Figure 2.1 as circles of different sizes (the actual orbits are neither circles nor spheres; they are much more complex patterns). Electrons have a negative charge that is equal in magnitude but opposite in sign to the positive charge of the proton. In its ideal state, an atom has an equal number of protons and electrons and thus is electrically neutral. Under certain circumstances, however, an atom may gain or lose an electron during a chemical reaction. Although atoms can gain or lose electrons, they never gain or lose protons or neutrons as a result of chemical processes.

An atom that loses or gains one or more electrons has a net electric charge and is called an **ion**. If the charge is positive, meaning that the atom has lost one or more electrons, the ion is called a *cation.* If the charge is negative, meaning that the atom has gained one or more electrons, the ion is called an *anion.* A convenient way to indicate ionic charges is to record them as superscripts. For example, Na^+ is the symbol for an atom of sodium that has given up an electron; Cl^- is the symbol for an atom of chlorine that has accepted an electron; and Fe^{2+} is the symbol for an atom of iron that has given up two electrons.

COMPOUNDS, MOLECULES, AND BONDING

Chemical **compounds** form when atoms of one or more elements combine with atoms of another element in a specific

| **compound** A combination of atoms of one or more elements in a specific ratio. |

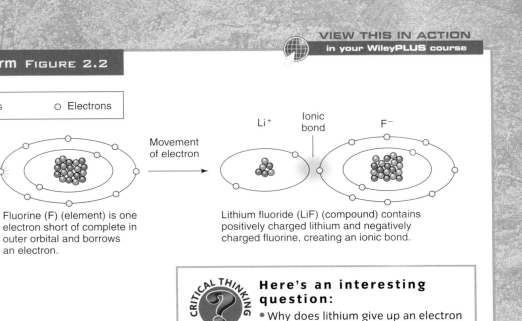

How ions and compounds form FIGURE 2.2

● Protons   ● Neutrons   ○ Electrons

Lithium (Li) (element) has one electron in outer orbital and donates an electron.

+

Fluorine (F) (element) is one electron short of complete in outer orbital and borrows an electron.

Movement of electron

Li⁺   Ionic bond   F⁻

Lithium fluoride (LiF) (compound) contains positively charged lithium and negatively charged fluorine, creating an ionic bond.

CRITICAL THINKING ?

Here's an interesting question:
• Why does lithium give up an electron when fluorine has so many more?

---

ratio. For example, sodium and chlorine combine to form sodium chloride (a mineral called *halite*, also known as table salt), which is written NaCl. For every Na atom in this compound, there is one Cl atom. The element that tends to form cations is written first, and the relative numbers of atoms are indicated by subscripts. For example, water forms when hydrogen (a cation, H^+) combines with oxygen (an anion, O^{2-}) in the ratio of two atoms of hydrogen to one atom of oxygen. Thus, for water, we write H_2O. FIGURE 2.2 shows how lithium and fluorine combine to form an ionic compound, lithium fluoride (LiF) that is used in the ceramics industry and for making lenses used for ultraviolet devices.

Properties of compounds differ from the properties of their constituent elements. For example, hydrogen and oxygen are both gases at Earth's surface, whereas water is a liquid. Similarly, sodium and chlorine are both highly toxic elements, whereas their compound, salt, is essential for life.

The smallest unit that has the properties of a given compound is a **molecule**. Do not confuse a molecule and an atom; the definitions are similar, but a molecular compound always consists of two or more atoms. Molecules are held together by electromagnetic forces known as **bonds**.

molecule The smallest chemical unit that has all the properties of a particular compound.

Bonding involves the transfer of electrons from one atom to another or, in some cases, the sharing of electrons. The four principal kinds of bonds are illustrated in FIGURE 2.3. You will see that different bonding explains the difference in properties of diamond and graphite.

bond The force that holds the atoms together in a chemical compound.

Why have we spent all this time learning about elements, compounds, and bonding? Because minerals are chemical compounds (or, in a few cases, simply chemical elements), the chemical elements and kinds of bonds determine the properties of a mineral, and minerals are the main building blocks of the solid Earth. Now let's look more closely at the characteristics that define minerals and help us to identify them.

CONCEPT CHECK   STOP

What requirements must be satisfied if a substance is to be called a mineral?

Why can water be separated into chemically distinct substances (hydrogen and oxygen), while gold cannot?

How does an atom differ from an ion?

What determines an element's atomic number?

What are the four main types of bonds in minerals?

Ionic bonding: When one atom transfers an electron to another, as illustrated in Figure 2.2, an attractive force is set up that creates an ionic bond. In a crystal such as table salt (NaCl), the sodium ions (red) are attracted to all of the neighboring chlorine ions (gray), not just to one of them. Thus the ionic bonds form a cubic lattice. Compounds with ionic bonds tend to have moderate strength and hardness.

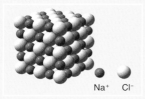

In table salt (sodium chloride, NaCl), each sodium cation is surrounded by chlorine anions.

Crystals of sodium chloride are rectangular, with straight edges.

Salt is a moderately hard solid that dissolves easily in water.

Covalent bonding: When electrons from different atoms "pair up," the force of this sharing is called a covalent bond. Note that electron sharing does not produce ions. These are the strongest chemical bonds, and elements and compounds with covalent bonds (such as diamond) tend to be strong and hard.

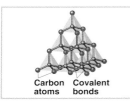

Diamond consists of carbon atoms connected in a network of covalent bonds. Each atom is connected to four others.

Diamond crystals appear in a rock called kimberlite. Covalent compounds are often strong and hard; diamond is one of the hardest substances known.

Cut and polished diamonds are prized gems. Tiny diamonds are used in industry for cutting and grinding instruments.

Metallic bonding: In metals, atoms are so tightly packed that electrons can be shared among several atoms. In fact, the outermost electrons are so loosely held that they can readily drift from one atom to another. This mobility of electrons explains why metals are good at conducting electricity and heat.

Atoms of gold are packed in the densest possible manner. Each atom is surrounded by, and in contact with, 12 other gold atoms

This nugget of gold was once embedded in rock, but weathering and erosion have removed most of the rock.

Gold is durable as well as malleable; it has been used as currency since ancient times. These are gold coins.

Van der Waals bonding: A weak attraction can occur between electrically neutral molecules that have an asymmetrical charge distribution. The positive end of one molecule will be attracted to the negative end of another molecule. For example, the carbon atoms in graphite form sheets in which each carbon atom has strong covalent bonds with three neighbors. The bonds between sheets are weak. This is why graphite feels slippery when you rub it between your fingers.

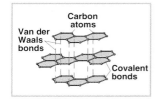

In graphite, carbon atoms form layers connected by covalent bonds. The layers are weakly held together by Van der Waals bonds.

Graphite is not a strong material and can be easily crumbled into small particles.

The "lead" in pencils is really graphite. When you write, the pressure of your hand breaks off a trail of carbon particles.

What Is a Mineral?

C hemists have been able to create millions of compounds in the laboratory, but there are only about 4000 compounds that qualify as minerals. It is important to keep the requirements for a substance to be a mineral clearly in mind. These requirements, as stated earlier, are that it be a *naturally formed solid*, be *formed inorganically*, and have a *specific chemical composition* and a *characteristic crystal structure*. Each of the items on this checklist is essential (**FIGURE 2.4**).

COMPOSITION OF MINERALS

An apparent confusion to the rule that a mineral must have a specific chemical composition is a phenomenon called *atomic substitution*. In some cases, two elements can be similar enough in size and in bonding properties that they can substitute for each other in a mineral. For example, magnesium and iron ions (Mg^{2+} and Fe^{2+}) are so similar in size, and have the same electrical charge, that one often takes the place of the other. The

Mineral or not a mineral? FIGURE 2.4

Ice is a mineral, though you may not usually think of it that way. It occurs in nature in the form of hexagonal crystals and has a specific chemical formula (H_2O).

Water is not a mineral, because it is not a solid. This criterion also means that such naturally occurring substances such as oil and natural gas cannot be considered minerals.

Bones are a tricky case. They do contain the same chemical compound found in a common mineral called apatite, but they are not minerals because they form by organic processes. Thus the bone in this modern crocodile skull is not a mineral.

However, this fossilized crocodile skull, from the Kenyan National Museum, is composed of minerals. Why? During fossilization, the original materials were replaced in an inorganic process called *mineralization*.

GEOGRAPHIC

mineral olivine (an important component of Earth's mantle) can occur as pure Fe_2SiO_4 or pure Mg_2SiO_4 or an intermediate mixture in which some of the Fe^{2+} cations are replaced by Mg^{2+} cations. We show atomic substitution in the chemical formula by putting parentheses around the substituting elements and a comma between them. The formula for olivine, therefore, becomes $(Mg, Fe)_2SiO_4$, indicating that Mg and Fe can substitute for one another in this mineral. Note that the ratio of cations to anions is not changed by atomic substitution, so the specific composition rule is not violated.

The composition requirements for minerals specifically rule out a material whose composition varies so much that it cannot be expressed by an exact chemical formula. An example of such a material is glass, which is a mixture of many elements and can have a wide range of compositions.

Glass—even naturally formed volcanic glass—also fails the test of having a characteristic **crystal structure**. In a *crystal*, the atoms are arranged in regular, repetitive geometric patterns, as shown in FIGURE 2.5 (on the following page). By contrast, the atoms in a liquid or in an amorphous solid such as glass are mixed up or randomly jumbled. Sometimes amorphous solids are referred to as *mineraloids*; an example of a mineraloid is opal, a familiar stone that is often used in jewelry.

> ■ **crystal structure**
> An arrangement of atoms or molecules into a regular geometric lattice. Materials that possess a crystal structure are said to be crystalline.

Coal fails the second of the four tests for a mineral because it is derived from the remains of plant material and was formed as a result of organic processes.

Steel (being made in the background) fails the first of the four tests for a mineral because it does not occur naturally. It is formed by extensive human processing of naturally occurring ores, which are minerals.

Quartz is an easily recognizable mineral. Its chemical formula is SiO_2. Note that some minerals have very complex formulas. For example, phlogopite, a form of mica, is $KMg_3AlSi_3O_{10}(OH)_2$. The important thing is that the elements combine in specific ratios.

Although **opals** are typically included in books about minerals, they are not true minerals because they do not have a specific composition and lack a crystalline structure. Opals are *mineraloids*.

▼ **A** The atoms of all crystalline materials are arranged in orderly lattices, like the cubical lattice illustrated here for a mineral called *galena* (PbS), the main source of lead. The atoms are so small that a cube of galena 1 cm on an edge would contain 10^{22} atoms (that's 1 followed by 22 zeros). The inset shows an exploded view of the packing arrangement of atoms in a galena crystal. The atoms are shown pulled apart along the black lines to demonstrate how they fit together. Compare the arrangement of atoms in NaCl (Figure 2.3); it is the same as the arrangement of lead and sulfur atoms in PbS.

All specimens of a given mineral have an identical crystal structure. Extremely sensitive scanning tunneling microscopes enable scientists to determine the crystal structures of minerals and actually detect the orderly arrangement of atoms in the mineral. As you can see in Figure 2.5, the atoms in a crystalline material resemble the regular, orderly rows in an egg carton.

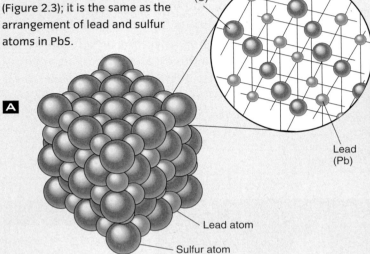

Sulfur (S)

Lead (Pb)

Lead atom

Sulfur atom

▼ **B** Atoms are too small to see with an optical microscope, but a scanning tunneling microscope can detect the location of the atoms in a crystal. This is an image of what such a microscope detected in a galena crystal; the sulfur atoms look like large bumps and the lead atoms like small ones.

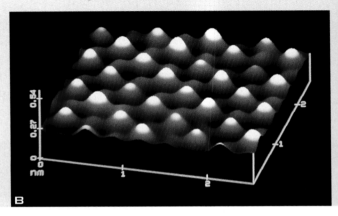

B

TELLING MINERALS APART

The compositions and crystal structures of minerals influence their physical properties and characteristics. If we have an unidentified mineral sample (such as **FIGURE 2.6**), we can apply a few simple tests to determine what mineral it is, without taking it to a laboratory or using expensive equipment. The properties most often used to identify minerals are the quality and intensity of light reflected from the mineral, crystal form and habit of the mineral, hardness, tendency to break in preferred directions, color, and specific gravity or density. Color, perhaps the most obvious characteristic, is often the least reliable identifier. Let's look at the properties that are used to identify different minerals.

Luster Suppose you collected a mineral sample like the one in Figure 2.6. What steps would you go through to identify it? One of the first things you would notice is

A "mystery" mineral FIGURE 2.6

Scientists have several effective, low-tech methods to identify minerals. We will use them to identify the mineral pictured here. First, note its metallic luster and cubic habit. (The answer is on page 48.)

how shiny it is, or what scientists call its **luster**. Different minerals can reflect light in different ways as well as with different intensities. The luster of the mystery sample in Figure 2.6 is *metallic*, meaning that it looks like a polished metal surface. Some other kinds of *nonmetallic* lusters you might encounter are *vitreous*, like that of glass; *resinous*, like that of resin; *pearly*, like that of pearl (**FIGURE 2.7**); or *greasy*, as if the surface were covered by a film of oil. Two minerals with almost identical color can have quite different luster.

Crystal faces and mineral habits

The ancient Greeks were fascinated by ice. They were intrigued by the fact that needles of ice are six-sided and have smooth, planar surfaces. The Greeks called ice *krystallos*. Eventually the word *crystal* came to be applied to

any solid body that has grown with flat or planar surfaces. The planar surfaces that bound a crystal are called *crystal faces*.

During the 17th century, scientists investigated crystal faces as a way to identify minerals. But the sizes of faces vary widely from one sample to another. Under some circumstances, a mineral species may grow a thin crystal; under others, the same mineral species may grow a fat crystal, as **FIGURE 2.8** shows. It is apparent from the figure that the overall crystal size and the relative sizes of crystal faces are not the same for these two crystals of quartz. In fact, crystal size and the relative sizes of crystal faces are not definitive for any mineral.

In 1669, a Danish physician, Nils Stensen (better known by his Latin name, Nicolaus Steno), unraveled the mystery of crystal faces. Steno demonstrated that the key property that identifies a given mineral is the angles between the faces. Steno's Law states that the angle between any corresponding pairs of crystal faces of a given mineral species is constant no matter what the overall shape or size of the crystal might be (see Figure 2.8).

Steno and other early scientists hypothesized that a mineral must have some kind of internal order that

Mineral luster FIGURE 2.7

Three examples of luster.

Vitreous
◀ Quartz (SiO_2) has a glassy luster.

Resinous
▶ Sphalerite (ZnS, a source of zinc) has a resinous luster, like dried tree resin.

Pearly
◀ Talc [$Mg_3Si_4O_{10}(OH)_2$] has a pearly luster.

Crystal faces and angles FIGURE 2.8

Crystals of the same mineral may differ widely in shape and size. However, the angles between faces will remain the same in all specimens. In the two quartz crystals pictured, numbers identify equivalent faces. According to Steno's Law, the angle between faces 1 and 3 (for example) is the same in both specimens.

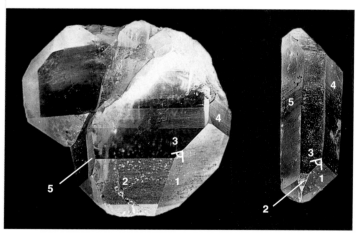

What Is a Mineral? 43

predisposes it to form crystals with constant interfacial angles. Support for their hypothesis finally arrived in 1912, when German scientist Max von Laue sent a beam of X-rays through a crystal and showed that the diffracted rays matched the patterns a geometric array would create. More recently, scanning tunneling microscopes have given more direct proof of atomic lattice structure, as shown in Figure 2.5.

Crystals develop planar faces most easily when mineral grains can grow freely in an open space. Because most mineral grains do not form in open, unobstructed spaces, nicely formed crystals are uncommon in nature, and very large crystals are very rare because large open spaces inside Earth are rare. Usually other mineral grains get in the way as minerals grow. As a result, most mineral grains have an irregular shape. However, in both a crystal and an irregularly shaped grain of the same mineral, all the atoms present are packed in the same strict geometric pattern. This is why we use the term *crystal structure*, rather than *crystal*, in the definition of a mineral.

Some minerals grow in such distinctive ways that their shape—called the mineral's **habit**—can be used as an identification tool. Our mystery sample in Figure 2.6 clearly has a *cubic* habit, because it looks like a collection of interlocked cubes. A very different example is the mineral chrysotile, shown in FIGURE 2.9,

Fibers of asbestos FIGURE 2.9

Some minerals have distinctive growth habits even though they do not develop well-formed crystal faces. The mineral chrysotile ($Mg_3Si_2O_5(OH)_4$) sometimes grows as fine, cotton-like threads that can be separated and woven into fireproof fabric. When the mineral occurs like this, it is said to have an *asbestiform* habit. Many different minerals can grow with asbestiform habits, and several are mined and commercially sold as asbestos.

which takes the form of fine fibers or threads. This *fibrous* habit is characteristic of asbestos minerals.

Hardness **Hardness**, like habit and crystal form, is governed by crystal structure and by the strength of the bonds between

habit The distinctive shape of a particular mineral.

hardness A mineral's resistance to scratching.

The Mohs Scale* of relative hardness of minerals TABLE 2.1

	Relative Hardness Number	Reference Mineral	Hardness of Common Objects
Softest	1	Talc	
	2	Gypsum	
	3	Calcite	Fingernail
	4	Fluorite	Copper penny
	5	Apatite	Pocketknife; glass
	6	Potassium feldspar	
	7	Quartz	
	8	Topaz	
	9	Corundum	
Hardest	10	Diamond	

The 10 minerals of the Mohs scale are shown above, starting with the softest, talc, in the upper left-hand corner, and proceeding across two rows to diamond, the hardest mineral in the lower right-hand corner.

*Named for Friedrich Mohs, a German mineralogist, who chose the 10 minerals of the scale.

SUMMARY

1 Minerals, Elements, and Compounds

1. Never confuse **rocks** and **minerals**. Minerals are to rocks as letters are to words. All specimens of a given mineral have the same basic composition. Rocks have variable compositions because they are aggregates of one or more kinds of minerals and may also contain fragments of other rocks, amorphous substances, and organic matter.

2. **Elements** are the most fundamental of all naturally occurring substances because they cannot be separated into chemically distinct materials. The minute particles that make up all matter, including elements, are **atoms**. All atoms of a particular element are the same.

3. Atoms are made of smaller particles called *protons*, *neutrons*, and *electrons*, which have no independent chemical

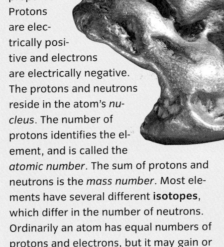

properties. Protons are electrically positive and electrons are electrically negative. The protons and neutrons reside in the atom's *nucleus*. The number of protons identifies the element, and is called the *atomic number*. The sum of protons and neutrons is the *mass number*. Most elements have several different **isotopes**, which differ in the number of neutrons. Ordinarily an atom has equal numbers of protons and electrons, but it may gain or lose electrons, in which case it becomes electrically charged and is called an **ion**.

4. **Compounds** consist of multiple elements, and therefore multiple types of

atoms. The smallest unit that has the properties of a given compound is a **molecule**. Molecules in a compound are held together by **bonds**. The most common form of bonding is *ionic bonding*, caused by the electrostatic attraction of two oppositely charged ions. Other forms of bonding are *covalent*, *metallic*, and *Van der Waals bonding*.

2 What Is a Mineral?

1. **Minerals** are *naturally occurring solids*, each with a unique **crystal structure**. Minerals are formed by *inorganic processes* and must have a *fixed chemical composition*. One extension of the last rule is *atomic substitution*, in which other atoms of like size and ionic charge may be substituted for specific atoms in a mineral without causing the crystal structure to change.

2. The *crystal faces* that bound a mineral are a direct consequence of its atomic lattice. In some cases, a mineral will not have enough room to grow identifiable crystals, but the underlying geometric lattice remains the same.

3. Several properties can be used to tell minerals apart. One of the most easily visible is a mineral's **luster**, or the qual-

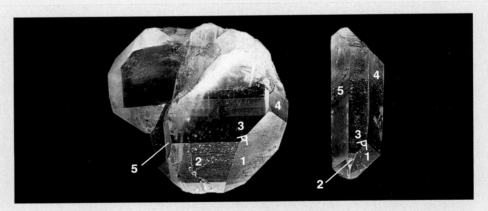

ity and intensity of light that reflects from it. Measuring the angles between adjacent faces is a useful way to identify a crystal, for though crystal size and overall shape in a mineral may vary, *Steno's Law* states that the angle between corresponding faces remains constant. Other factors that can aid in mineral identification include the distinct external shape, or **habit**; a mineral's resistance to

scratching, known as its **hardness**; how a mineral behaves when broken, known as **cleavage**; and its **density**, or *specific gravity*. Color may also be useful but is often misleading. However, when a mineral is rubbed on a *streak plate*, it produces a thin layer of powdered mineral, known as its **streak**, which is more reliable than color for identification.

3 Mineral Families

1. The distribution of elements in Earth's crust is far from even. Only 12 elements are present at a level of more than one part in a thousand by mass, and of these, oxygen and silicon dominate. As a result, the number of naturally occurring minerals is relatively small, and the number of important rock-forming minerals is even smaller—only about 30 or so.

2. **Silicate minerals**, based on the $(SiO_4)^{4-}$ anion, are the most abundant family of **rock-forming minerals**. They can adopt a variety of crystal structures because of the ability of the *silicate anions* to link together. These structures include chains, sheets, and three-dimensional lattices. **Oxide minerals** are the next most abundant family. Other important mineral families include *sulfides* and *sulfates*, *carbonates*, and *phosphates*. Less abundant minerals are called **accessory minerals**, and they usually do not affect the properties of the rock they are found in. Nevertheless, these minerals are frequently of great economic importance as sources of metal ore.

3. Two minerals can have the same chemical formula but different crystal structures. These different forms are *polymorphs*. Other minerals occur in nature as *native elements* that are uncombined with other elements.

4. Silicates are the most common minerals. Large, complex ions can be formed by *polymerization*, a process of bonding based on the sharing of an oxygen atom. Three factors affect the identity of a silicate mineral: whether the silicate tetrahedra are single or polymerized, which cations are present, and how the cations are distributed.

4 Mineral Resources

1. All of the metals used in industry and elsewhere come from minerals mined on Earth. The distribution of **ore deposits** around the world is very uneven, and as a result no country is self-sufficient in all the mineral resources needed.

2. Mining disturbs the ground surface. Responsible mining restores the surface and/or mitigates the impact after mining is finished, but small-scale mining, which is widespread, does not restore the surface and, in addition, is all too often the source of releases of toxic chemicals, such as mercury, into the environment.

KEY TERMS

- **mineral** p. 36
- **rock** p. 36
- **element** p. 36
- **atom** p. 36
- **isotopes** p. 37
- **compound** p. 37
- **molecule** p. 38
- **bond** p. 38
- **crystal structure** p. 41
- **luster** p. 43
- **habit** p. 44
- **hardness** p. 44
- **cleavage** p. 45
- **streak** p. 46
- **density** p. 48
- **ore deposit** p. 53

Rocks: A First Look

Rocks, as we saw in Chapter 2, are different from minerals. The important distinction is that rocks are *aggregates*. This means that rocks are collections of mineral particles (and sometimes other types of particles such as organic debris or bits of volcanic glass) stuck together or intergrown to make a coherent mass. Rocks usually consist of several types of minerals, but sometimes they are made of just one common mineral, such as quartz or calcite. In any case, a rock will

◀ **Igneous rock**

This rounded granite boulder on Mt. Desert Island in Acadia National Park, Maine, was transported to its current location by a glacier. The boulder is sitting on a thick platform of granite that was once part of an enormous chamber full of molten rock—magma—underlying a volcano. The volcano is now gone and its top has eroded away, leaving the solidified remnants of the magma chamber exposed. The weathered surfaces of these granites make them appear buff-colored.

▲ Granite is an igneous rock. It forms from magma that cools and solidifies slowly, deep in the crust. The minerals in this sample are quartz (gray), potassium feldspar (light pink), plagioclase feldspar (white), and biotite (black). The same minerals are present in the sample of gneiss, shown on the facing page, but the overall appearance of the two rocks is quite different.

always contain many grains of the constituent mineral or minerals.

The kind of rock that forms depends on the environment in which it forms, and for that reason, rocks are the record keepers of Earth's long history. Rocks are the words that tell the story; minerals are the letters that form the words.

THE THREE ROCK FAMILIES

Rocks are grouped into three large families, according to the processes that formed them (see FIGURE 3.1).

Within each of these families, called *igneous, sedimentary,* and *metamorphic* rock, there is a range of possible *mineral assemblages*—the types and relative proportions of the minerals that constitute the rock—and a range of possible *textures*—the overall appearance of a rock because of the size, shape, and arrangement of mineral grains. Mineral assemblages and textures help identify rocks (for example, whether it is granite or basalt), textures help separate the rock families (for example, whether a sample is igneous or metamorphic), and the two together reveal much about the particular environment in which a given rock formed.

Visualizing
The three rock families FIGURE 3.1

Landscapes reflect the kind of rock that underlies them, just as mineral assemblages and textures serve to define rock types (see insets next to scenery photographs).

Metamorphic rock ▶

The beautiful Lauterbrunnen Valley in Switzerland lies within a great mountain range—the Alps—where the rocks have been uplifted and chemically and physically altered by enormous tectonic forces. In this photo, the Lauterbrunnen Falls cascade down a steep cliff of metamorphic rock.

◀ Sedimentary rock

This remarkable landscape is part of Bryce Canyon National Park in Utah. The horizontal rock layers are mainly sandstones and limestones. The tall spires, called *hoodoos*, and the deeply incised crevices between them, are the result of erosion over many, many years. Bryce Canyon today is a desert, but the grains of sediment in these rocks were originally deposited in an environment that was alternately hot and dry (sandstone) or covered by a shallow sea (limestone).

◀ This photo shows a sample of gneiss (pronounced "nice"), a metamorphic rock similar to the rocks that form many of the cliffs in the Swiss Alps. This rock may have started as either a sedimentary or igneous rock, but it has been altered by heat and pressure so that a new set of minerals has developed, along with its banded appearance. The minerals present are potassium feldspar (light pink), plagioclase (white), quartz (gray), and biotite (black). The minerals are similar to those in the igneous rock granite, shown on the facing page, but the overall appearance is quite different.

◀ Sandstone is a sedimentary rock that consists largely of grains of quartz, with traces of other minerals. The grains became rounded as they were transported by running water to the place where they were deposited. After being deposited in layers, the grains eventually became cemented together and turned into sandstone.

Actual inset samples are 7 cm wide and 4 cm high.

THE ROCK CYCLE

Earth's surface is a meeting place. It is where the activities of Earth's internally driven processes—the movement of continents as a result of plate tectonics, earthquakes, elevation of mountains, eruptions of volcanoes—confront the quicker-paced activity of Earth's surface layers, the atmosphere, hydrosphere, and biosphere, which constantly break down and erode away the surface rocks. The surface layers are the most dynamic parts of the Earth system. The external forces of wind, water, ice, and life constantly modify the surface, breaking down and cutting away material here, depositing material there, and sculpting the landscapes that surround us. These external forces are the most obvious ones of the Earth system; they are the forces that confront us every day in our daily lives. We are less aware of Earth's internal forces because they act more slowly and to a large extent are neither seen nor felt. But the internal forces play an important role in the Earth system because they cause changes such as the elevation of mountain ranges and the slow shifting of continents, and those changes in turn cause climates to change and influence the places where plants and animals can live.

All of this activity, both fast and slow, is called the **rock cycle**, one of the great cycles that drive the Earth system, as shown in **FIGURE 3.2**. The Earth system has three great cycles: the *rock cycle*, the *hydrologic cycle*, and the *tectonic cycle*. The three cycles are interconnected, and they influence everything that happens on and in Earth. Earth would not be a habitable planet without the three great cycles. We will discuss the cycles and the ways they interact at many places in later chapters.

The rock cycle has no beginning or ending: instead, it is an endless process, powered by Earth's internal heat energy and by the incoming energy from the Sun. We will discuss specific details of the rock cycle at several places in later chapters. For example, in Chapter 4, we will discuss the processes involved in **weathering**, by which bedrock breaks into smaller rock

> **rock cycle** The set of crustal processes that form new rock, modify it, transport it, and break it down.

> **weathering** The chemical and physical breakdown of rock exposed to air, moisture, and living organisms.

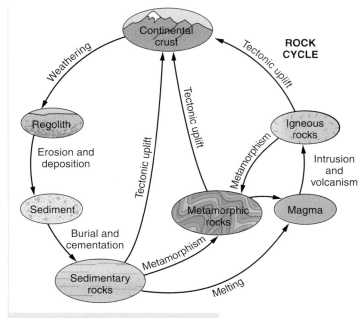

ROCK CYCLE

The rock cycle FIGURE 3.2

The cycle of rock change has been active since our planet became solid and internally stable. It continually forms and re-forms rocks of the three major families. Not even the most ancient igneous and metamorphic rocks that we have found are the original rocks of Earth's crust, for they were recycled eons ago.

and mineral fragments, and **erosion**, the group of related processes by which the products of weathering are moved around on Earth's surface. Weathering and erosion are critical steps in the rock cycle; they are powered by energy from the Sun. Earth's internal heat energy powers the forces that thrust up mountains and cause volcanoes to erupt, processes that expose new rocks to the forces of weathering and erosion.

> **erosion** The wearing away of bedrock and transport of loosened particles by a fluid, such as water.

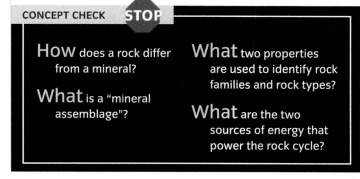

CONCEPT CHECK STOP

How does a rock differ from a mineral?

What is a "mineral assemblage"?

What two properties are used to identify rock families and rock types?

What are the two sources of energy that power the rock cycle?

Igneous Rocks

LEARNING OBJECTIVES

Distinguish between volcanic and plutonic igneous rocks.

Identify different textures of igneous rocks and explain the physical processes that produce them.

Identify the three main types of volcanic rock and their plutonic equivalents.

I gneous rocks (named from the Latin *ignis*, meaning *fire*) form by the cooling and *crystallization* of **magma**. When magma cools, mineral grains begin to crystallize, just as ice crystals form in water as it cools.

> **igneous rocks**
> Rocks that form by cooling and solidification of molten rock.

> **magma** Molten rock.

The physical properties of the rock will differ, depending on whether the cooling process is slow or fast. The *rate of cooling* determines how large the individual mineral grains in the rock will grow—slow cooling yields large grains; fast cooling yields tiny grains. Grain size determines the appearance or *texture* of a rock. The *composition* of the magma determines the final mineral assemblage in the solidified rock. Let's examine each of these factors more closely because mineral assemblage and texture are the properties by which igneous rocks are classified and named.

RATE OF COOLING

Even with a cursory look at igneous rocks, you can see that some contain easily visible mineral grains and others do not. Rate of cooling is the main factor that distinguishes the two groups. **Volcanic rocks** (also called *extrusive igneous rocks*) form at Earth's surface, where lava contacts the much cooler air and water. They solidify so quickly that large mineral grains have no time to form.

> **volcanic rock** An igneous rock formed from lava.

> **plutonic rock** An igneous rock formed underground from magma.

Plutonic rocks (also called *intrusive igneous rocks*) form when magma crystallizes deep underground. This process is much slower and therefore gives the mineral grains time to grow larger.

Rapid cooling: Volcanic rocks and their textures

Sometimes lava cools so rapidly that mineral grains do not have a chance to form at all. The resulting volcanic rock is not crystalline but *glassy* (**FIGURE 3.3A**).

Volcanic rock textures FIGURE 3.3

A *Glassy texture*. This photograph shows obsidian obelisks from a Mayan grave in Guatemala. Though these were sculpted by humans, a shiny, curvy appearance is typical for volcanic glass.

B *Aphanitic texture*. In this fine-grained rock, individual mineral grains cannot be discerned with the naked eye. Note also the vesicles (due to trapped volcanic gas). This sample of basalt is from Hawaii, and each of the largest vesicles is about the size of a small pea. Such a bubble-rich rock is called *vesicular* basalt.

C *Porphyritic texture*. This volcanic rock from Nevada contains large mineral grains, or *phenocrysts,* suspended in an aphanitic material called the *groundmass*. The largest grains visible in the photo are approximately 6 mm in length.

Putting Rocks Under a Microscope

Earth scientists can identify the minerals in an aphanitic rock by studying it under a microscope.

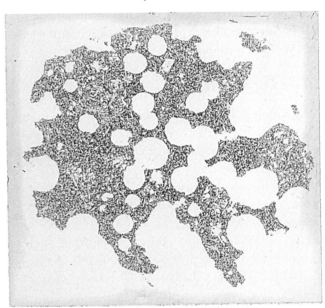

A The specimen of aphanitic volcanic rock from Figure 3.3 has been cut and polished into a wafer that is thin enough to allow light to pass through. Earth scientists call these rock slices *thin sections*. This magnified thin section is about 0.03 mm thick, as thin as a piece of tissue paper, and about 3.5 cm across.

B Using polarized light, Earth scientists can see individual mineral grains and identify them by optical and other properties. Using a microscope to study thin sections is a common technique in Earth science to analyze rocks of all types for many purposes.

CRITICAL THINKING ?

Here's an interesting question:
• There are two obviously different minerals visible in photo B. One is olivine, the other is feldspar. One appears yellowish, the other white. Which is which?

More often, mineral grains do form in solidifying lavas, but they are extremely small, and can be seen only under magnification (see *What an Earth Scientist Sees*). Rocks with this very fine-grained texture are said to be *aphanitic* (FIGURE 3.3B).

Most igneous rocks contain mineral grains of approximately the same size—that is, they are unimodal. But some volcanic rocks are bimodal in grain size and have a texture known as *porphyritic* (FIGURE 3.3C), which means they consist of large mineral grains embedded in an aphanitic matrix. This happens when magma starts to crystallize slowly at depth and grows large grains before it erupts, and the suspended large grains erupt along with the liquid lava. Once erupted, the remaining magma cools rapidly to form an aphanitic matrix in which are embedded the large mineral grains that formed earlier.

Dissolved gases, too, can affect the texture of volcanic rock. Erupting lava may froth and bubble as dissolved gases are released; if the froth is blasted into the air and cools quickly, it forms *pumice*, which is a glassy mass of bubbles. As lava cools, the viscosity increases and it becomes increasingly difficult for gas bubbles to escape. When the lava finally solidifies into rock, the last bubbles to form may become trapped; the bubble holes are called *vesicles*. In basaltic lava, this process can create a volcanic rock with lots of bubble holes that looks like Swiss cheese (see Figure 3.3B).

Slow cooling: Plutonic rocks and their textures

Unlike the minerals in volcanic rocks, those in plutonic rocks usually have time to form mineral grains that can be readily seen by the unaided eye. This coarse-grained texture is said to be *phaneritic* (FIGURE 3.4). Exceptionally large mineral grains (sometimes up to several meters!) typically form in the last stage of crystallization of a plutonic rock, when gases build up in the remaining magma. The vapor facilitates the growth of large crystals, because chemicals can migrate quickly to the growing faces. A coarse-grained plutonic rock with mineral grains larger than 2 cm in diameter is called a *pegmatite*.

The process that produces porphyritic texture in volcanic rocks also occurs in plutonic rocks. Magma starts to slowly crystallize deep underground, forming big crystals; then the remaining magma, with large crystals suspended within it, is intruded higher in the crust—though not extruded as a lava. In the new position, cooling is more rapid, so the remaining magma crystallizes to smaller crystals. The resulting texture is porphyritic even though the matrix is phaneritic.

CHEMICAL COMPOSITION

Scientists subdivide the most common igneous rocks into three broad categories based on their silica contents. Rocks that contain large amounts of silica (about 70% SiO_2 by weight) are usually light-colored. They are said to be *felsic* (a word formed from *feldspar* and *silica*), because feldspar and quartz are the most common minerals found in them. At the other end of the scale are rocks called *mafic* (a word formed from *magnesium* and *ferrous*, or iron-rich), which contain large amounts of dark-colored minerals, such as olivine and pyroxene, rich in magnesium and iron. They are usually lower in silica content (about 50% SiO_2 by weight). Mafic rocks that contain even less silica, and consist almost entirely of olivine and pyroxene, are said to be *ultramafic*. Igneous rocks with silica contents about 60% SiO_2 by weight, halfway between felsic and mafic, are said to be *intermediate* in composition. The origins of different magma types, and the way magmas crystallize into different kinds of igneous rocks through processes such as Bowen's Reaction Series, are discussed in Chapter 9.

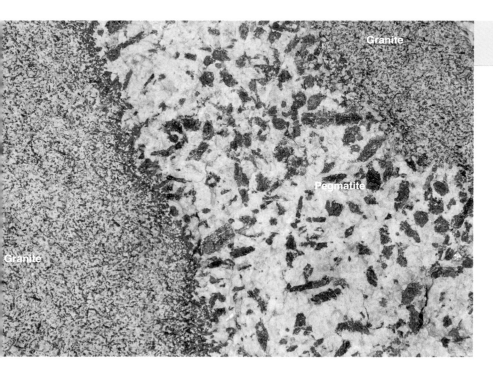

Plutonic rock and textures
FIGURE 3.4

Two distinct grain sizes of plutonic rock can be seen in this granite specimen from California's Sierra Nevada. Both textures are phaneritic; in the two outside layers there are small but visible grains of plagioclase, potassium feldspar, quartz (white), and biotite (black). Sandwiched between them is a vein of pegmatite, which contains the same minerals but in much larger grains.

Scientists organize igneous rock types on the basis of both grain size and silica content. The results are shown in **TABLE 3.1**. The rocks on the left are volcanic (aphanitic and fine-grained), and the ones on the right are plutonic (phaneritic and coarse-grained). The ones on the top contain the most silica and are the lightest in color (that is, they are felsic), while those on the bottom contain the least silica and are darkest in color (they are mafic).

Volcanic and plutonic equivalents TABLE 3.1

Grain Size →

Silica Content of Magma / **Silica Content** ↑

Silica Content of Magma	Resulting Volcanic Rocks		Resulting Plutonic Rocks	
High (= 70%–75%)	**Rhyolite** lies at the felsic, high-silica end of the scale, and consists largely of quartz and feldspars. It is usually pale, ranging from nearly white to shades of gray, yellow, red, or lavender.		**Granite**, the plutonic equivalent of rhyolite, is common because felsic magmas usually crystallize before they reach the surface. It is found most often in the continental crust, especially the cores of mountain ranges.	
Intermediate (= 60%)	**Andesite** is an intermediate-silica rock, with lots of feldspar mixed with darker mafic minerals such as amphibole or pyroxene. It is usually light to dark gray, purple, or green.		**Diorite** is the plutonic equivalent of andesite, an intermediate-silica rock.	
Low (= 45%–50%)	**Basalt**, a mafic rock, is dominant in oceanic crust, and the most common igneous rock on Earth. Large, low-viscosity lava flows from shield volcanos and fissures are usually basaltic. Dark-colored pyroxene and olivine give it a dark gray, dark green, or black color.		**Gabbro** is the plutonic equivalent of basalt, a low-silica rock.	

BIOGENIC SEDIMENTS AND BIOGENIC SEDIMENTARY ROCK

Biogenic sediment is composed of the remains of plants and animals. This includes the hard parts of large animals, such as shells, bones, and teeth, as well as fragments of plant matter, such as wood, roots, and leaves (see **FIGURE 3.8B**). Two of the most common rock types that come from biogenic sediments are limestone, formed of the calcium carbonate skeletons of marine invertebrates, and coal, formed from partially decomposed terrestrial plant material. (Note that limestone can be either biogenic or chemical in origin.)

Limestone is the most abundant biogenic sedimentary rock. It is formed from the lithified shells and other skeletal material from marine organisms. Some of these organisms build their shells or skeletons from calcite, but most construct them from aragonite, which, like calcite, is calcium carbonate. During diagenesis the aragonite is transformed to calcite, which becomes the main mineral of limestone.

Calcite is sometimes replaced by the mineral dolomite (a carbonate mineral containing both magnesium and calcium); the resulting rock is called *dolostone.* Another kind of biogenic rock, which consists of extremely tiny particles of quartz, is called *chert.* The quartz in chert does not come from sand, but from the shells of microscopic sea animals.

An important class of biogenic sediment consists of the accumulated remains of terrestrial plants. Over time, and with pressure, this mass gradually becomes **peat.** Eventually, given enough time and pressure, peat may lithify and become **coal.** Lithification (also called *coalification*) involves further compaction, release of water, and slow chemical changes that weld the plant fragments together, thereby making the coal relatively lower in water and richer in carbon than the original peat.

ROCK BEDS

When you look at an outcrop of sedimentary rock, such as the one shown in **FIGURE 3.9**, one of the first things you will notice is the **bedding.** The banded appearance comes from the fact that sedimentary particles are laid down in distinct *beds,* or *strata.* Over time, the mineral composition of the sediments in a particular location may change, or they may be transported or deposited in different ways. This will cause the adjacent strata to look different. The boundary between adjacent strata is called a *bedding surface.* It is the presence of bedding and bedding surfaces that indicates that a rock was once sediment.

INTERPRETING ENVIRONMENTAL CLUES

Just as history books record the changing patterns of civilization, sedimentary rocks record the environmental history of our planet. Layers of sedimentary rock, like pages in a book, record how environmental conditions have changed throughout Earth history. Earth scientists can "read" this story by interpreting the evidence in the rocks.

> **biogenic sediment** Sediment that is composed primarily of plant and animal remains, or precipitates as a result of biologic processes.

> **limestone** A sedimentary rock that consists primarily of the mineral calcite.

> **peat** A biogenic sediment formed from the accumulation and compaction of plant remains.

> **coal** A combustible rock formed from the lithification of plant-rich sediment.

> **bedding** The layered arrangement of strata in a body of sediment or sedimentary rock.

Layers of rock: Bedding FIGURE 3.9

The Bungle Bungle Range in northwestern Australia derives its unique coloration from layers of sandstone that have different permeabilities. Algae grow in the more permeable strata, tinting the rock black.

Patterns formed by currents of water or air moving across sediment can be preserved and later exposed on bedding surfaces. For example, bodies of sand that are being moved by wind, streams, or coastal waves are often rippled; these may be preserved in sandstone as *ripple marks* (see FIGURE 3.10A and B). Similarly, *mud cracks* (FIGURE 3.10C and D), fossil tracks (see FIGURE 3.11), and even raindrop impacts can be recorded on bedding surfaces, attesting to moist surface conditions at the time they were formed.

Ancient and modern features compared FIGURE 3.10

▲ A Ripples are forming in shallow water near the shore of Ocracoke Island, North Carolina.

▲ B Almost identical ripples are exposed on a bedding surface of sandstone at Artists Point, Colorado National Park, Colorado.

▲ C Mud cracks formed on this modern riverbed as the river dried up.

▲ D Similarly shaped mud cracks are preserved on the surface of shale exposed at Ausable Chasm, New York. We can infer that this rock formation was deposited in an intermittently wet environment, such as a seasonal lakebed or a tidal flat. Note that in both (B) and (D) the present-day environment bears no relation to the environment in which the sedimentary rocks were deposited.

Rocks as Resources

 hen we say *resources*, people immediately tend to think of gold, diamonds, and other things of high value. But the most valuable resources are rather mundane things such as crushed rock for road construction and gypsum for making sheetrock walls. In the following short section we briefly review some of the ways we use rocks in our society.

ROCKS AS CONSTRUCTION MATERIALS

Almost all kinds of rocks find a use somewhere in the construction industry, but some rocks have more desirable properties than others and so are more widely used. Most of us are familiar with beautiful stone countertops in kitchens and bathrooms—the materials used tend to be limestone, marble, or phaneritic igneous rock, and they are popular because they have striking patterns and colors. But the rocks differ in durability. The main mineral in limestone and marble is calcite, which is soft and scratches easily, and calcite reacts with many common household liquids, so staining and marking are problems. Igneous rocks contain silicate minerals such as quartz, feldspar, and mica, which are physically durable and chemically resistant and do not get easily scratched or stained.

Limestones, marbles, and igneous rocks—especially granites—are widely used as cut stone for building, although the cost tends to restrict their use to publicly funded structures. Granite is far more durable than either limestone or marble because calcite is soluble, albeit slowly, in rainwater (see **FIGURE 3.21**). High-grade metamorphic rocks have many of the same physical properties as granites and can be used for the same purposes in the construction industry. Low-grade metamorphic rocks, especially slates, which break into flat sheets due to slaty cleavage, are widely used for durable—though expensive—roof coverings.

The amount of rock used for countertops and buildings pales to insignificance compared to the use of crushed rock for road construction and for aggregate in concrete and asphalt. The most suitable types of rock for crushed rock are basalt, limestone, and gneiss; their use amounts to nearly 10 tons per year for every person living in the United States.

Effects of rainwater on limestone FIGURE 3.21

Statues carved from limestone several hundred years ago, on the façade of the Bayeux Cathedral, Normandy, France, show the disfiguring effects caused by the acidity of rainwater.

Hydrothermal gold deposit FIGURE 3.22

Delicate sheets of native gold were deposited in the center of a quartz vein in Burgin Hill Mine, California.

MINERAL RESOURCES IN IGNEOUS ROCKS

Many valuable ore minerals occur as accessory minerals in igneous rocks. On occasion, the amount of a mineral present is sufficient to warrant mining. Granites, for example, and especially pegmatitic granites, sometimes contain beryl, one of the sources of beryllium, a metal important in the nuclear power industry. Other minerals found in granites are the source for the metals tantalum, tin, and niobium, each of which can be added to metals such as copper and iron to improve strength and lithium, which is used in ceramics and in batteries for watches and hearing aids.

MINERAL RESOURCES IN CHEMICAL AND BIOGENIC SEDIMENTARY ROCKS

Chemical and biogenic sedimentary rocks, because of the way they form, are concentrations of substances precipitated from sea or lake water. Many of the precipitates have valuable chemical properties. For example, the fertilizer element, potassium, is recovered from marine evaporites, and phosphorus, the most important of all fertilizer elements, is found in rocks of marine biogenic origin. Other examples are halite (table salt) and gypsum (for sheetrock), both of which are found in marine evaporites. One especially important chemical sedimentary rock is banded iron formation, discussed earlier in this chapter (see *What an Earth Scientist Sees: A Change in the Atmosphere*, page 76).

GOLD IN METAMORPHIC ROCKS

Metamorphic rocks start out as sedimentary or igneous rocks, so any mineral deposit present in those rocks will be preserved as a metamorphosed mineral deposit. There are many such metamorphosed deposits in terrains of geologically older rocks that have been subjected to metamorphism.

As discussed earlier in this chapter, metamorphism is accompanied by the expulsion of aqueous fluids, and in many cases the fluids flow through fractures and form veins that are predominantly quartz-rich (see Figure 3.15). Metamorphic fluids are never pure water—small amounts of many minerals are present in solution and these may be deposited with the quartz. When gold is among the minerals deposited, valuable ore deposits can be the result. Some of the gold produced in California is thought to have its origins in metamorphic vein deposits (FIGURE 3.22).

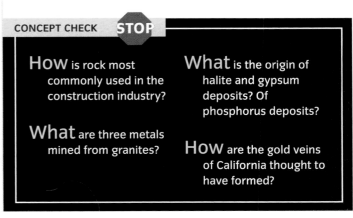

CONCEPT CHECK STOP

How is rock most commonly used in the construction industry?

What are three metals mined from granites?

What is the origin of halite and gypsum deposits? Of phosphorus deposits?

How are the gold veins of California thought to have formed?

Amazing Places: The Navajo Sandstone

NATIONAL GEOGRAPHIC

Global Locator

Zion National Park

A

B

C

D

Eolian Transport

Sand Sea

Stream

Transport

Appalachians   Grenville Mountains

Many of the rock formations in Utah's Zion National Park resemble sand dunes—because that is what they originally were. The imposing cliffs (A), the roadside hill (B), and the stunningly etched ground seen from above (C) were all once part of a vast sea of sand, larger and thicker than the Sahara Desert is today. The dunes lithified into sandstone and subsequently eroded into the formations you see today. The sedimentary formation they belong to, called the *Navajo Sandstone,* extends over several states in the American Southwest and attains a thickness of 700 m in Zion National Park.

But if all of this rock was once sand, that poses a puzzle: Where did all the sand come from? The dune patterns indicate that the prevailing winds came from the north, but no mountain ranges of suitable size or age can be found there. The mountain range we call the Rockies did not yet exist in the Lower Jurassic Period, 190 million years ago, when the Navajo sand was being deposited.

However, the Appalachian Mountains did exist then. Lifted up by a plate tectonic collision that began 500 million years ago, they were once as tall as the Himalaya are today and extended all the way into what is now Texas. By the Lower Jurassic the Appalachians had eroded considerably—and scientists now think that it was their sediment that formed the Navajo Sandstone. Rivers flowing westward carried the erosional sediment toward what is today the center of the continent (D). As the climate became arid, winds transported the now-dried river sands southward, where they became one of the largest sand seas that has ever existed on Earth.

SUMMARY

1 Rocks: A First Look

1. Rocks are coherent aggregates, or complex *assemblages of minerals*, that are sometimes mixed with other materials, such as volcanic glass and organic matter.

2. Rocks come in three major families. **Igneous rocks** are formed by the solidification of **magma**. **Sedimentary rocks** are formed by the **deposition** of many layers of sediment. **Metamorphic rocks** start as either sedimentary or igneous rocks but change their form and mineral assemblage as a result of high temperature, high pressure, or both.

3. The **rock cycle** consists of all the processes whereby rocks within and on top of Earth's crust are **weathered**, and the products of weathering are **eroded**, transported, deposited as sediment, converted into new rock, metamorphosed, melted, uplifted, and exposed again to weathering.

2 Igneous Rocks

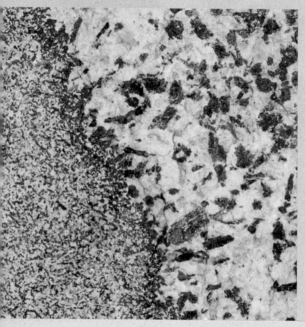

1. The process of solidification of cooling magma is called *crystallization.* The process of crystallization of magma influences the final properties of igneous rock, such as its *texture* and grain size. The crystallization process and the rock's final *mineral assemblage* in turn depend on such factors as the magma composition and the rate of cooling. A rock formed by rapid cooling of magma contains very small mineral grains.

2. Igneous rocks can be classified into **volcanic** and **plutonic rocks**. Volcanic rocks crystallize from lava; plutonic rocks crystallize underground from magma. Because lavas cool rapidly, mineral grains in volcanic rocks are small, microscopic, or even absent. These rocks are known as *aphanitic*. Plutonic rocks, on the other hand, are *phaneritic,* and have easily recognized mineral grains.

3. *Felsic* rocks are typically light in color, have high silica contents, and consist largely of quartz and feldspar. *Mafic* rocks are typically darker in color, with lower silica content, and consist largely of iron-containing minerals such as olivine and pyroxene. The most common volcanic rock is *basalt,* a dark, low-silica, mafic rock. *Granite* is a light-colored, high-silica, felsic rock that is a dominant constituent of the continental crust.

3 Sedimentary Rocks

1. Sediments and sedimentary rocks are our best record keepers of how climates and environments have changed throughout Earth's history. Earth scientists use three broad categories to distinguish types of sediment—*clastic*, *chemical*, and *biogenic*.

2. **Clastic sediment** consists of fragmented rock and mineral debris produced by weathering, together with broken remains of organisms. The fragments, called *clasts*, are classified on the basis of size. Clasts may become *sorted* during transport by water or wind but typically remain unsorted when transported by glaciers. Volcaniclastic sediments, also called *pyroclasts*, are volcanic in origin.

3. Clastic sedimentary rocks, like sediments, are classified mainly on the basis of clast size. **Conglomerate**, **sandstone**, and *mudstone* are the common rock equivalents of gravel, sand, and mud respectively. **Shale** is a mudstone that readily splits into thin layers.

4. **Lithification** is the group of processes that transform loose sediment into sedimentary rock. The principal processes are *compaction* of sediment, *cementation*, and *recrystallization* of mineral grains.

5. **Chemical sediment** is formed when substances carried in solution in lake water or seawater are precipitated, generally as a result of evaporation or other process that concentrates dissolved substances.

6. **Biogenic sediment** is composed of the accumulated remains of organisms. Plants, animals, and microscopic life-forms may all contribute skeletal and/or organic material to sediments.

7. When sediment is turned into sedimentary rock, the **bedding**, or layered arrangement of beds, is generally preserved. The presence of bedding and *bedding surfaces*, the boundaries between adjacent beds, indicates that the rock was once sediment.

8. Chemical sedimentary rocks formed by evaporation are **evaporites**. *Banded iron formations*, though uncommon, are significant both economically and for understanding the history of our planet. **Limestone**, **peat**, and **coal** are important kinds of biogenic sedimentary rocks.

9. A *sedimentary facies* is a single unit consisting of sedimentary rocks with the same composition, deposited at more or less the same time. Earth scientists learn about the history of a region by studying the way that different sedimentary facies adjoin and grade into each other in space and time.

4 Metamorphic Rocks

1. New rock textures and new mineral assemblages develop when rocks are subjected to elevated temperatures and stresses. **Metamorphism** is a term that describes all such processes that occur at higher temperatures and stresses.

2. Metamorphism commences at a temperature of about 150°C, which typically occurs at a depth of about 5 km. Between 5 and 15 km depth, rocks are subjected to *low-grade* metamorphism. The region of *high-grade* metamorphism lies from 15 km to the depth where melting commences, typically between 30 and 40 km.

3. *Pore* fluids enhance metamorphism by permitting material to dissolve, move around, and be precipitated somewhere else in the rock. *Veins* in metamorphic rocks mark the passageways through which pore fluids escaped from a rock undergoing metamorphism.

4. *Differential stress* during metamorphism produces a distinctive texture known as **foliation** marked by parallel cleavage planes. It is particularly noticeable in rocks containing minerals of the *mica* family. Foliation developed in low-grade metamorphic rocks is known as *slaty cleavage*. In higher-grade metamorphic rocks foliation is called *schistosity*.

5. The names of metamorphic rocks are based partly on texture and partly on composition. Foliated metamorphic rocks developed from shale are **slate**, *phyllite*, **schist**, and **gneiss** in order of increasing metamorphic grade. **Marble**, the metamorphic product of limestone, and **quartzite**, the metamorphic product of sandstone, are two common kinds of metamorphic rock that lack cleavage. Both rocks are *monomineralic*, or very nearly so.

5 Rocks as Resources

1. Essentially all kinds of rocks can be used somewhere in the construction industry. Limestone, marble, and granite are widely used as cut stones for building. The principal materials used as crushed rock for road building and concrete aggregate are basalt, limestone, and gneiss.

2. Igneous rocks are the source of many valuable ore minerals. Examples of metals recovered from igneous rocks are beryllium, tantalum, and lithium.

3. Chemical sedimentary rocks are the source of much of the world's potassium for fertilizers, table salt, and gypsum for sheetrock. Biogenic rocks are the source of phosphorus minerals used for fertilizers.

KEY TERMS

1. When a volcano erupts, spewing forth a column of hot volcanic particles, the particles are tiny fragments of solidified magma that slowly fall down to Earth's surface, forming a layer of sediment. Would the rock formed from cemented volcanic particles be igneous or sedimentary? Can you think of other circumstances that might form rocks that are intermediate between two of the major rock families?

2. Volcanic rock is sometimes called *extrusive* and plutonic rock is sometimes called *intrusive*. Why do you think that Earth scientists describe them this way? (You may wish to look up these words in a dictionary.)

3. Do any sedimentary rocks outcrop in the area where you live? If so, see if you can recognize the kinds of rocks present and identify the environment in which the sediments were deposited.

4. Exploration for oil has led to the discovery of about 14 km of sedimentary rock on the continental shelf of eastern North America. The oldest beds were deposited in the Jurassic Period, 150 million years ago. The youngest are still being deposited today. What is the average rate of deposition?

5. Examining the texture of a rock is an important rule-of-thumb for Earth scientists because the texture can reveal whether the rock has been metamorphosed and under what conditions. Explain why. Would this rule work for both foliated and nonfoliated rocks? Why (or why not)?

What is happening in this picture ?

This roof in Ireland was made from a commonly occurring planar metamorphic rock.

- What kind of rock do you think it is?

- What properties of this rock would make it very suitable for use as a roof tiling material?

- Do you think this rock is used in modern roofing? Why, or why not?

1. On this illustration, locate and label the following processes of the rock cycle:

   weathering   melting
   tectonic uplift   metamorphism
   burial and cementation   erosion and deposition
   intrusion and volcanism

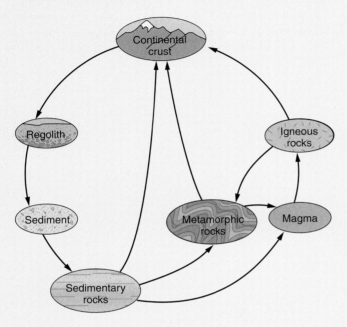

2. The three major rock families are _____.
   a. igneous, sedentary, metamorphic
   b. sedimentary, metamorphosis, igneous
   c. sedimentary, igneous, metamorphic
   d. metamorphic, igneous, metamorphosis

3. These two photographs are closeup views of two igneous rocks. Label each appropriately with the following terms:

   porphyritic-texture volcanic rock

   phaneritic-texture plutonic rock

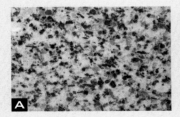 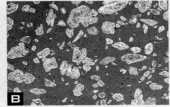

4. Which of the samples depicted in the photographs in question 3 records two distinct phases of cooling—slow followed by rapid?
   a. The sample labeled "A."
   b. The sample labeled "B."
   c. Neither. Both rocks cooled at the same rate.

5. Which of the following are plutonic igneous rocks correctly paired with their compositionally equivalent volcanic igneous rocks:
   a. granite/rhyolite; gabbro/andesite; diorite/basalt
   b. granite/basalt; gabbro/rhyolite; diorite/andesite
   c. diorite/basalt; granite/rhyolite; gabbro/andesite
   d. granite/rhyolite; diorite/andesite; gabbro/basalt

6. _____ sediment forms from loose rock and mineral debris produced by weathering and erosion.
   a. Clastic   b. Biogenic   c. Chemical

7. Which of the three rock samples in these photographs is not a sedimentary rock?
   a. Sample A
   b. Sample B
   c. Sample C
   d. All of the rock samples displayed are sedimentary rocks.

8. Which of the following **sedimentary** rocks is typically composed entirely of calcite?

a. sandstone
b. limestone
c. shale
d. conglomerate
e. evaporite

9. Which of the following is a medium-grained sedimentary rock of clastic origin?

a. sandstone
b. limestone
c. shale
d. conglomerate
e. evaporite

10. Which of the following rocks is most likely to have formed in a seasonally dry desert lake?

a. sandstone
b. limestone
c. shale
d. conglomerate
e. evaporite

11. Metamorphism will occur under relatively _____ conditions.

a. high temperature and high pressure
b. high temperature and low pressure
c. low temperature and high pressure
d. All of the above statements are correct.

12. Pore fluids enhance metamorphism by _____.

a. permitting material to dissolve
b. permitting chemical components to move around and be precipitated somewhere else
c. speeding up some chemical reactions
d. All of the above statements are correct.

13. Foliated metamorphic rocks derived from shales are _____, in order of increasing metamorphic grade.

a. phyllite, slate, and schist or gneiss
b. phyllite, gneiss, and schist or slate
c. slate, phyllite, and schist or gneiss
d. slate, gneiss, and phyllite or schist

14. The two rock samples shown in these photographs are nonfoliated metamorphic rocks. Sample A is composed of calcite, and Sample B was formed through the metamorphism of sandstone. Which of the following correctly identifies the two samples?

a. Sample A is quartzite and Sample B is schist.
b. Sample A is quartzite and Sample B is marble.
c. Sample A is marble and Sample B is quartzite.

15. Limestone is not the most desirable stone for buildings because _____.

a. it is fissile and tends to splinter
b. it is too hard and cannot be cut easily
c. it slowly dissolves in rainwater
d. All of the above statements are true.

Weathering, Soils, and Mass Wasting

4

In May 2003, New Hampshire's beloved landmark, the Old Man of the Mountain, fell victim to the forces of nature. The Old Man was granitic rock, sculpted in part by glacial activity during the last ice age, that bore a conspicuous resemblance to a human face jutting out from a cliff—a sort of Mt. Rushmore sculpted by nature. It had been commemorated just three years earlier on the New Hampshire state quarter (see inset). Caretaker Niels Nielson and his son are making a regular inspection of the monument.

The Old Man's demise can be attributed to the inexorable forces of weathering. Mere rock cannot resist the combined assault of water, ice, chemical changes, and gravity. Deep inside the Old Man's granite features, water repeatedly froze and thawed; each freeze widened and extended numerous minute cracks inside the rock, while at the same time chemical changes chiseled away at the outer edges. After hundreds of years of this wear and tear, the five granite ledges that created the man's profile broke apart and tumbled to the bottom of the hill.

New Englanders were dismayed at the loss of the seemingly timeless monument. Plans are underway to create a museum with high-tech imagery of the formation before its fall.

In reality, rocks exposed at Earth's surface are never timeless or immutable. They are constantly broken down by weathering and erosion; without these processes, we would not have soil or beaches. At the same time, new rock is constantly forming deep in Earth's crust. This constant recycling of material is the rock cycle, discussed at the beginning of Chapter 3.

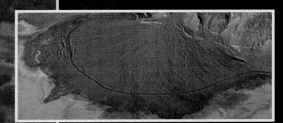

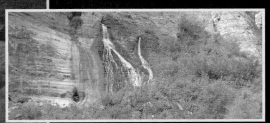

The Hydrologic Cycle

Define and describe the hydrologic cycle.

Identify the main pathways in the hydrologic cycle.

Identify the main reservoirs in the hydrologic cycle.

Four great reservoirs make up the Earth system: the lithosphere, hydrosphere, atmosphere, and biosphere. Water moves—and helps move materials—among all four spheres. Water evaporates from the ocean, and water vapor enters the atmosphere. Water is a constituent of many common minerals (such as micas and clays) in the lithosphere, where it is tightly bonded in their crystal structures. And, of course, water is a fundamental component of living things in the biosphere. The **hydrologic cycle**, also called the *water cycle*, describes how water moves among these four reservoirs (**FIGURE 5.1**), and the scientific study of water is called *hydrology*.

> **hydrologic cycle**
> A model that describes the movement of water through the reservoirs of the Earth system; the *water cycle*.

Process Diagram

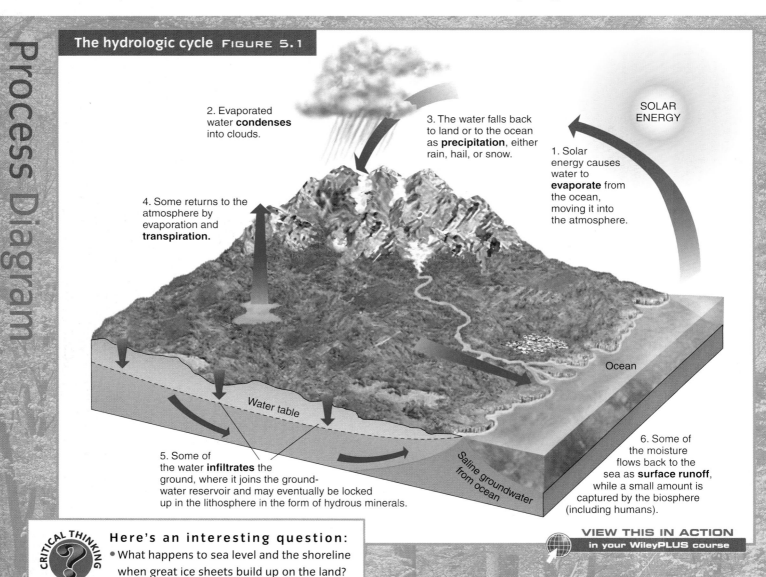

The hydrologic cycle FIGURE 5.1

SOLAR ENERGY

1. Solar energy causes water to **evaporate** from the ocean, moving it into the atmosphere.

2. Evaporated water **condenses** into clouds.

3. The water falls back to land or to the ocean as **precipitation**, either rain, hail, or snow.

4. Some returns to the atmosphere by evaporation and **transpiration.**

5. Some of the water **infiltrates** the ground, where it joins the ground-water reservoir and may eventually be locked up in the lithosphere in the form of hydrous minerals.

6. Some of the moisture flows back to the sea as **surface runoff**, while a small amount is captured by the biosphere (including humans).

Water table

Saline groundwater from ocean

Ocean

CRITICAL THINKING

Here's an interesting question:
• What happens to sea level and the shoreline when great ice sheets build up on the land?

VIEW THIS IN ACTION
in your WileyPLUS course

The hydrologic cycle as part of the Earth system FIGURE 5.2

Diagram of the hydrologic cycle, showing how water moves between the reservoirs to maintain a balanced cycle. Because the amount of water in the cycle is fixed, when any part of the cycle changes, the rest of the cycle adjusts so that a new balance is reached.

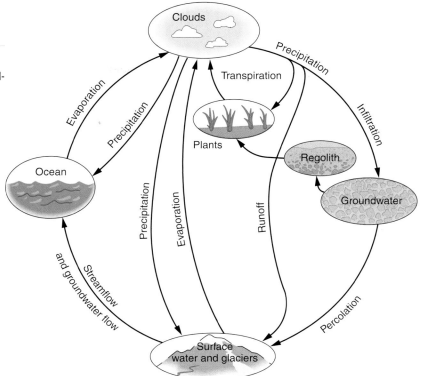

HYDROLOGIC CYCLE

---

evaporation The process by which water changes from a liquid to a vapor.

transpiration The process by which water taken up by plants passes directly into the atmosphere.

condensation The process by which water changes from vapor into a liquid.

deposition The process by which water changes from a vapor into a solid.

Water moves through the hydrologic cycle along numerous pathways and processes. These include **evaporation** and **transpiration**, both of which are powered by energy from the Sun. Depending on local conditions of temperature, pressure, and humidity, some of the water vapor in the atmosphere will undergo **condensation**, changing to a liquid, or **deposition**, changing to a solid and falling back to the land or ocean as rain, snow, or hail via the process of **precipitation**. Some of this precipitation becomes **surface runoff**, whereas some trickles directly into the ground via **infiltration**.

WATER IN THE EARTH SYSTEM

The schematic representation of the hydrologic cycle (see **FIGURE 5.2**) is a *closed cycle* of *open systems*. Because it is a closed cycle, the total amount of water is fixed. However, all of the local reservoirs within the cycle, such as rivers and trees, are free to gain or lose water. They sometimes do so quite dramatically, as during a flood or drought.

The water cycle is easily observable and readily studied. We can measure the amount of global precipitation; using satellite monitoring, we can even measure the amount of evaporation. With these

precipitation The process by which water that has condensed in the atmosphere falls back to the surface as rain, snow, or hail.

surface runoff Precipitation that drains over the land or in stream channels.

infiltration The process by which water works its way into the ground through small openings in the soil.

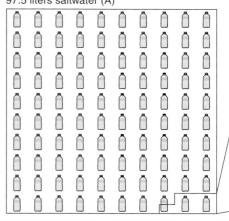

The world's water resources (in proportion):
97.5 liters saltwater (A)

1.85 liters frozen water (B)

0.64 liters groundwater (C)

10 milliliters surface water (D)*
(= 2 teaspoons)
*includes water in biosphere and atmosphere

The vast majority of Earth's water is salty (A), frozen (B), or underground (C). The most visible everyday sources of fresh water, such as rivers, lakes, and the atmosphere (D), together comprise less than one hundredth of a percent of Earth's water budget.

Oxbow lake After the cutoff, silt and sand seal the abandoned channel, producing a lake. The lake fills with fine sediment and eventually turns into a swamp.

Floodplain The meandering river channel dominates the floodplain. Parts of the channel were abandoned after cutoff events, when the river cut across a meander loop to create a shorter, more direct path.

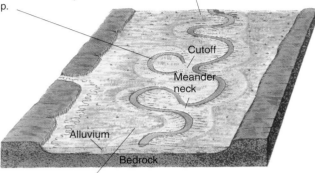

Natural levees are created during flooding, when sand and silt are deposited next to the channel creating belts of higher land on either side of the channel.

▲ **A** Some of the landforms created by sediment in stream valleys are floodplains, oxbow lakes, natural levees, and alluvial fans.

B Where this stream emerges from the mountains into Death Valley, California, it abruptly slows and deposits its sediment load. This has created a symmetrical alluvial fan, covered by a braided system of channels. The stream was dry at the time the photograph was taken.

C Where the Nile River empties into the Mediterranean Sea, the sediment it deposits has formed a fan-shaped delta that supports the green vegetation seen in this satellite image. Its triangular shape, similar to the Greek letter delta (δ), gave rise to the term *delta* that we use today.

DECONSTRUCTING A COAST

1839

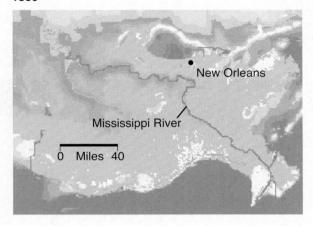

New Orleans

Mississippi River

0   Miles   40

1993   ☐   Land loss on coast

New Orleans

2090 *Projection*

New Orleans

Disappearing coastline FIGURE 5.7

The Mississippi River has been changed by levees and channels, slowing deposition of new sediment in its delta. As a result, the Mississippi Delta has been shrinking. Marshes give way to open water and barrier islands shrink. The top image shows the delta in 1839, the middle image in 1993, and the bottom image shows a projection in 2090. The degradation of the delta is partly responsible for the vulnerability of New Orleans to hurricane damage.

Prominent deltas do not form in places where strong wave, current, or tidal action redistributes sediment as quickly as it reaches the coast. However, if the rate of sediment supply exceeds the rate of coastal erosion, a delta will form. The converse holds, too: If the rate of deposition slows down, the delta will disappear (see FIGURE 5.7).

LARGE-SCALE TOPOGRAPHY OF STREAM SYSTEMS

Streams are governed by a simple principle: Water flows downhill. Rainwater that falls on the land surface will move from higher to lower elevations under the influence of gravity. A stream's headwater region is the area of relatively higher elevation from which streams have their source. Small, high-gradient *tributary streams* carry water downslope from the headwater region, combining their flow to form a larger stream. The gradient gradually decreases toward the low-lying region of the stream's mouth.

Every stream is surrounded by its **drainage basin** (sometimes called a *catchment* or a *watershed*). Drainage basins range in size from less than a square kilometer to areas the size of subcontinents. In general, the greater a stream's annual discharge, the larger its drainage basin. The vast drainage basin of the Mississippi River encompasses more than 40% of the total area of the contiguous United States (see FIGURE 5.8). From an environmental perspective, a drainage basin is a more natural geographic entity than a country or a state, because issues of water supply, pollution,

> **drainage basin**
> The total area from which water flows into a stream.

Another aspect of flood prediction is the real-time monitoring of storms and water levels. Scientists can combine information about the weather with their knowledge of a river basin's characteristics and topography to forecast the peak height of a flood and the time when the crest will pass a particular location. Such forecasts, which are often made with the aid of computer models and Geographic Information Systems (GISs), can be very useful for planning evacuation or defensive measures.

Understandably, many people throughout history have been unsatisfied with simply predicting floods and have attempted to prevent them. River channels are often modified or engineered for the purpose of flood control and protection as well as to increase access to floodplain lands, facilitate transport, enhance drainage, and control erosion. The modifications usually consist of some combination of widening, deepening, straightening, clearing, or lining of the natural channel. All of these approaches are collectively called *channelization.*

Like dams, channelization projects can contribute to the economic well-being of a community, but at a price. Channel modifications interfere with natural habitats and ecosystems. The aesthetic value of the river can be degraded, and water pollution aggravated. Projects sometimes control flooding in the immediate area but contribute to more intense flooding downstream. Perhaps most importantly, any modification of a channel's course or cross-section renders invalid the hydrologic data collected there in the past. During the Mississippi River floods of 1973 and 1993, experts could not account for water levels that were higher than predicted by the historical data; the likely cause was extensive upstream modifications of the river channel by humans.

SURFACE WATER RESOURCES

A reliable water supply is critical—not only for human survival and health, but also for the role it plays in industry, agriculture, and other economic activities (see FIGURE 5.14 on the following page). Twenty-six countries worldwide, with a total population of almost 250 million people, are today designated as water-scarce. The lack of water in these countries places seri-

ous constraints on agricultural production, economic development, health, and environmental protection.

Globally, crop irrigation accounts for about 73% of the demand for water, industry for about 21%, and domestic use for the remaining 6%, though the proportions vary from one region to another. Demand in each of these sectors has more than quadrupled since 1950. Population growth is partly responsible for the increasing demand, but improvements in standards of living around the world have also contributed to the large increase in water use per capita over the past few decades. The total amount of water being withdrawn (that is, diverted from rivers, lakes, and groundwater) for worldwide human use is now about eight times the annual streamflow of the Mississippi River.

Sometimes, because of population growth and development, regions with the greatest demand for water do not have an abundant and readily available supply of surface water. For this reason, surface water is often transferred from one drainage basin to another, sometimes over long distances. Besides raising political issues related to water rights, such *interbasin transfer* can have negative environmental impacts (see the *Case Study* on page 152).

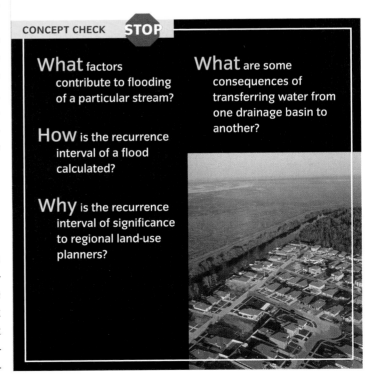

CONCEPT CHECK STOP

What factors contribute to flooding of a particular stream?

How is the recurrence interval of a flood calculated?

Why is the recurrence interval of significance to regional land-use planners?

What are some consequences of transferring water from one drainage basin to another?

WATER ACCESS AND USE

In arid and poorly developed regions, large proportions of inhabitants lack clean water supplies. Most water is consumed by agriculture for irrigation.

▲ Water gatherers

Women wait to fill water jugs from an irrigation canal in southern Ethiopia. In this drought-plagued area, more than 80% of rural inhabitants lack access to clean drinking water.

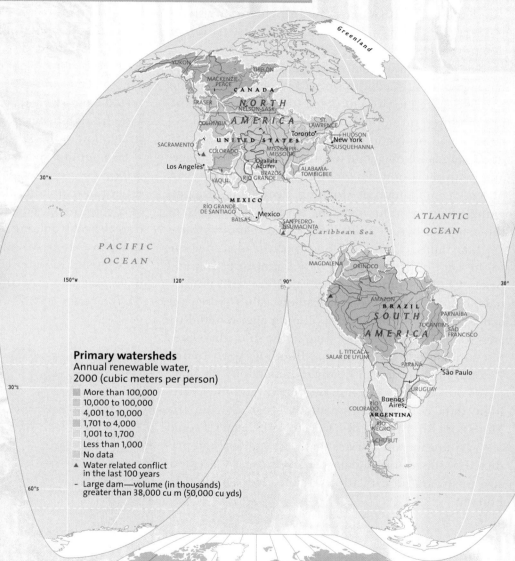

Primary watersheds
Annual renewable water, 2000 (cubic meters per person)

- More than 100,000
- 10,000 to 100,000
- 4,001 to 10,000
- 1,701 to 4,000
- 1,001 to 1,700
- Less than 1,000
- No data
- ▲ Water related conflict in the last 100 years
- − Large dam—volume (in thousands) greater than 38,000 cu m (50,000 cu yds)

Access to fresh water ▶

In many regions, drinkable water is becoming scarce because of increasing demand and decreasing quality. Pollution of surface and ground water with contaminants and organisms is a major threat. Contamination of aquifers by pesticides and heavy metals is especially worrisome.

Percent of total population using improved drinking water sources, 2000

- More than 90%
- 76% – 90%
- 51% – 75%
- 25% – 50%
- Less than 25%
- No data available

WATER AVAILABILITY

Within a watershed, water availability depends on both precipitation and the number of people the water must support.

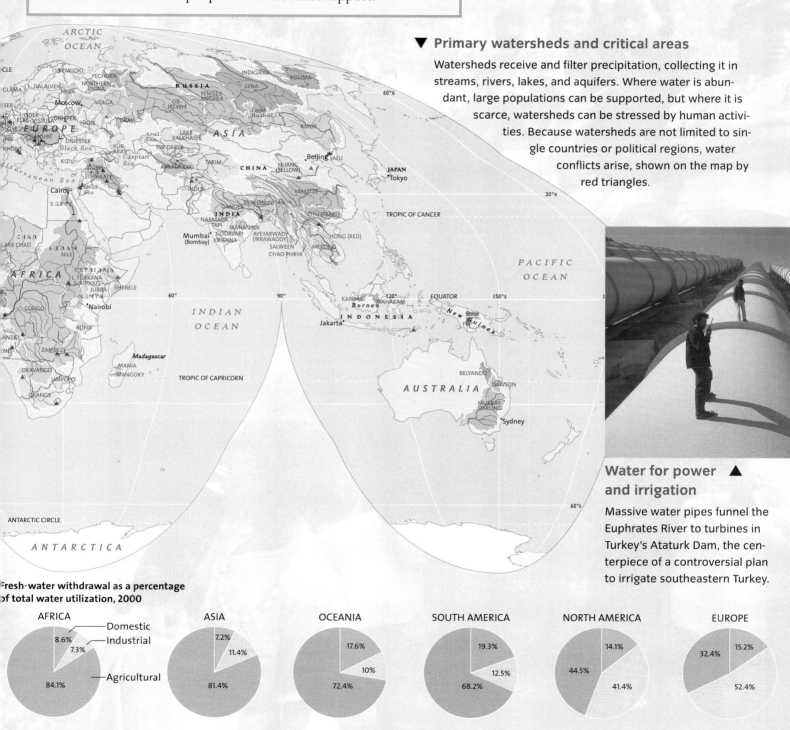

▼ Primary watersheds and critical areas

Watersheds receive and filter precipitation, collecting it in streams, rivers, lakes, and aquifers. Where water is abundant, large populations can be supported, but where it is scarce, watersheds can be stressed by human activities. Because watersheds are not limited to single countries or political regions, water conflicts arise, shown on the map by red triangles.

Water for power ▲ and irrigation

Massive water pipes funnel the Euphrates River to turbines in Turkey's Ataturk Dam, the centerpiece of a controversial plan to irrigate southeastern Turkey.

Fresh-water withdrawal as a percentage of total water utilization, 2000

AFRICA	ASIA	OCEANIA	SOUTH AMERICA	NORTH AMERICA	EUROPE
Domestic 8.6%	7.2%	17.6%	19.3%	14.1%	15.2%
Industrial 7.3%	11.4%	10%	12.5%	44.5%	32.4%
Agricultural 84.1%	81.4%	72.4%	68.2%	41.4%	52.4%

Mono Lake

California's Mono Lake, in the Sierra Nevada Mountains, has been the site of a collision between human water needs and the needs of a unique habitat and its wildlife.

These calcium carbonate spires (see **A**) were once underwater. But in 1941, the Los Angeles Aqueduct began diverting water from four of the six streams that empty into Mono Lake. With insufficient input to make up for its evaporation losses, the lake shrank to half its original volume and doubled in salinity. Migratory birds, such as the nation's second largest colony of California gulls, were placed in jeopardy. Their food supply (brine shrimp) was dying because of the high salinity of the water, and the islands on which they nested were in danger of being connected to the mainland because of the retreating water, making the birds vulnerable to predators.

By the late 1980s, the lake and its ecosystem were nearing collapse. In 1994, after a court battle between the City of Los Angeles and environmental groups, both sides agreed to a plan to raise the water level by 5 m and compensate Los Angeles for the loss of part of its water supply.

Water levels in the lake have risen about halfway to the target set by the agreement, and the lake is expected to reach its target level in 15 to 20 years. Dams

that once diverted streams into the aqueduct (see **B**) have been reengineered to do the opposite: They maintain a steady flow into Mono Lake and divert water into the aqueduct only if there is an overflow. For the California gulls (see **C**), the change may have come just in time.

Fresh Water Underground

LEARNING OBJECTIVES

Define water table.

Explain how porosity and permeability of rocks affect the motion of groundwater.

Identify two types of aquifers.

Explain why some wells require pumping, while others flow unaided.

Describe how subsidence relates to groundwater.

Describe how a cave forms.

L ess than 1% of all water in the hydrosphere cycle is **groundwater**. Although this sounds small, the volume of groundwater is 40 times greater than that of all the water in fresh-water lakes and streams. Water can be found everywhere beneath the land surface, even beneath parched deserts. About half of it is near the surface, no more than 750 m below ground. At greater depths, the pressure exerted by overlying rocks reduces the pore space, making it difficult for water to flow freely. In addition, water that occurs at great depths tends to be *briny*, or rich in dissolved mineral salts, and not well suited for human use. Therefore, from a practical perspective, we can think of groundwater as the water found between the land surface and a depth of about 750 m, even though

groundwater Subsurface water contained in pore spaces in regolith and bedrock.

water table The top surface of the saturated zone.

an equally large amount of water is present at greater depths.

THE WATER TABLE

Much of what we know about groundwater has been learned from the accumulated experience of generations of people who have dug or drilled millions of wells. This experience tells us that a hole penetrating the ground ordinarily first encounters a zone in which the spaces between the grains in regolith or bedrock are filled mainly with air, although the material may be moist to the touch. This is the *zone of aeration*, also known as the *vadose zone* (see **FIGURE 5.15**).

After passing through the zone of aeration, the hole reaches the **water table** and enters the *saturated zone,*

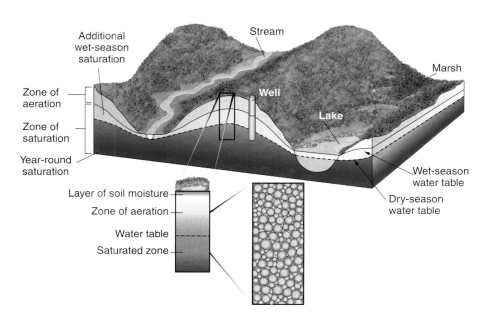

Additional wet-season saturation

Stream

Zone of aeration

Zone of saturation

Year-round saturation

Well

Marsh

Lake

Wet-season water table

Dry-season water table

Layer of soil moisture

Zone of aeration

Water table

Saturated zone

Water under the ground
FIGURE 5.15

A well first passes through the zone of aeration, where pores in the soil are filled with both air and water. Eventually, it reaches the water table, where the pore spaces are completely filled with water. Underground water exists everywhere, and surface water occurs wherever the ground intersects the water table.

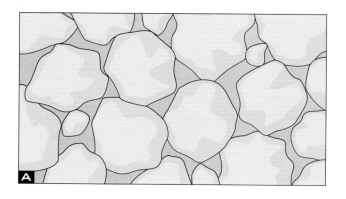

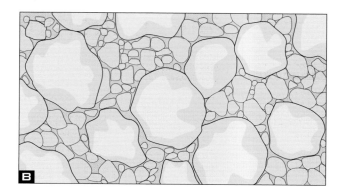

Pore space

Cement

Porosity in sediments and rocks

FIGURE 5.16

In these examples, all of the pore spaces are filled with water, as they would be in the saturated zone. **A** The porosity is about 30% in this sediment, with particles of uniform size. **B** This sediment, in which fine grains fill the space between larger grains, has a lower porosity, around 15%. **C** In sedimentary rock, the porosity may be reduced by cement that binds the grains together and fills the pores.

0   0.5mm

also known as the *phreatic zone,* in which all openings are filled with water. The water table is high beneath hills and low beneath valleys. This may seem surprising, because the surface of a glass of water or a lake is always level. But water underground flows very slowly and is strongly influenced by surface topography. If all rainfall were to cease, the water table would slowly flatten. Seepage of water into the ground would diminish and then stop entirely, and streams would dry up as the water table fell. During droughts, the depression of the water table is evident from the drying up of springs, streambeds, and wells. Repeated rainfall, which soaks the ground with fresh supplies of water, maintains the water table at a normal level and keeps surface water bodies replenished.

Whether it is deep or shallow, the water table marks the upper limit of readily usable groundwater. For this reason, a major aim of groundwater specialists and well drillers is to determine the depth and shape of the water table. To do this they must first understand how groundwater moves and what forces control its distribution underground.

HOW GROUNDWATER MOVES

Most groundwater is in motion. Unlike the swift flow of rivers, however, which is measured in kilometers per hour, the movement of groundwater is so slow that it is measured in centimeters per day or meters per year. The reason is simple: Whereas the water of a stream flows through an open channel, groundwater must move through small, constricted passages. Therefore, the rate of groundwater flow is dependent on the nature of the rock or sediment through which the water moves, especially its porosity and permeability.

Porosity and permeability

Porosity determines the amount of fluid a sediment or rock can contain. The porosity of sediment is affected by the size and shape of the particles and the compactness of their arrangement (see FIGURE 5.16A and B) The porosity of a sedimentary rock

porosity The percentage of the total volume of a body of rock or regolith that consists of open space.

is also affected by the extent to which the pores have been filled with cement (see FIGURE 5.16C). Plutonic igneous rocks and metamorphic rocks, which consist of many closely interlocked crystals, generally have lower porosities than do sediment and sedimentary rocks. However, joints and fractures may increase their porosity.

A rock with low porosity is likely also to have low **permeability**. However, high porosity does not necessarily mean high permeability, because both the sizes and the continuity of the pores (that is, the extent to which the pores are interconnected) influence the ability of fluids to flow through the material.

> **permeability** A measure of how easily a solid allows fluids to pass through it.

> **percolation** The process by which groundwater seeps downward and flows under the influence of gravity.

Percolation

After water from a rain shower soaks into the ground, or infiltrates, some of it evaporates, while some is taken up by plants. The remaining water continues to **percolate** under the influence of gravity until it reaches the water table. The "perc test" that must be carried out when a new septic system is being installed is a measure of percolation. The movement of groundwater in the saturated zone is similar to the flow of water that occurs when you gently squeeze a water-soaked sponge. Water moves slowly through very small pores along threadlike paths. The water flows from areas where the water table is high toward areas where it is lower. In other words, it generally flows toward surface streams or lakes (see FIGURE 5.17). Some of the flow paths turn upward and enter the stream or lake from beneath, seemingly defying gravity. This upward flow occurs because groundwater is under greater pressure beneath a hill than beneath a stream or lake. Because water tends to flow toward points where pressure is low, it flows toward bodies of water at the surface.

Recharge and discharge

Recharge of groundwater occurs when rainfall and snowmelt infiltrate the ground and percolate downward to the saturated zone (see Figure 5.17). The water then moves slowly along its flow path toward zones where **discharge** occurs. (Note that earlier in the chapter we used the term *discharge* for a somewhat different concept—the flow of water along a stream channel.) In discharge zones, subsurface water either

> **recharge** Replenishment of groundwater.

> **discharge** The process by which subsurface water leaves the saturated zone and becomes surface water.

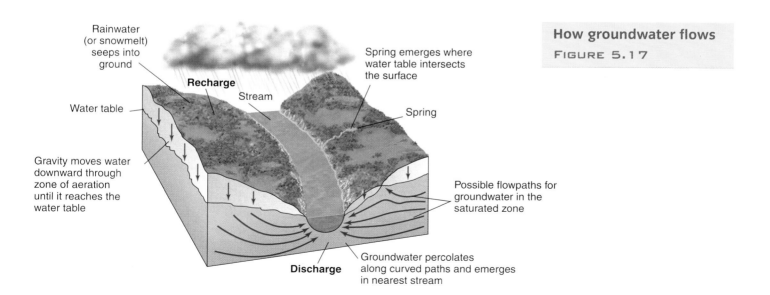

Rainwater (or snowmelt) seeps into ground

Water table

Gravity moves water downward through zone of aeration until it reaches the water table

Recharge

Stream

Spring emerges where water table intersects the surface

Spring

Possible flowpaths for groundwater in the saturated zone

Discharge

Groundwater percolates along curved paths and emerges in nearest stream

How groundwater flows
FIGURE 5.17

flows out onto the ground surface as a **spring** or joins bodies of water such as streams, lakes, ponds, swamps, or the ocean. The discharge of groundwater is what maintains the base flow of a stream. Pumping groundwater from a well also creates a point of discharge. The amount of time water takes to move through the ground to a discharge area depends on distance and rate of flow. It may take as little as a few days or as long as thousands of years.

WHERE GROUNDWATER IS STORED

When we wish to find a reliable supply of groundwater, we search for an **aquifer** (Latin for "water carrier"). An aquifer is not a body of water—it is a body of rock or regolith that is water-saturated, as well as porous and permeable. Gravel and sand generally make good aquifers; sandstone is often a good aquifer, as is fractured or cavernous limestone or granite. An aquifer in which the water is free to rise to its natural level is called an *unconfined aquifer* (see **FIGURE 5.18**). In a well drilled into an unconfined aquifer, the water will rise to the level of the surrounding water table. To bring it to the surface one would need a pump or a bucket.

A *confined aquifer* is overlain by impermeable rock units, called *confining layers*, or **aquicludes** (see **FIGURE 5.19**). Common aquicludes are shale and clay layers. The water in a confined aquifer is held in place by the overlying aquiclude, and its recharge zone may be many kilometers away at a higher elevation. If a well is drilled into the aquifer, the high water pressure due to the elevation of the recharge zone will cause the water to rise or even flow out of the well without having to

Aquifers, confined and unconfined FIGURE 5.18

An unconfined aquifer is open to the atmosphere through pores in the rock and soil above the aquifer. In contrast, the water in a confined aquifer is trapped between impermeable rock layers.

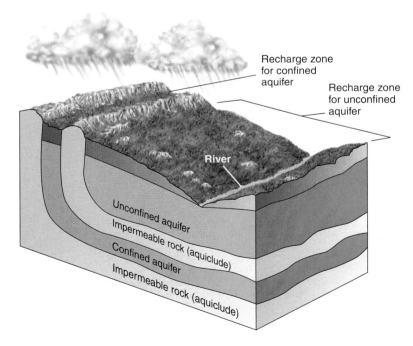

Recharge zone for confined aquifer

Recharge zone for unconfined aquifer

River

Unconfined aquifer

Impermeable rock (aquiclude)

Confined aquifer

Impermeable rock (aquiclude)

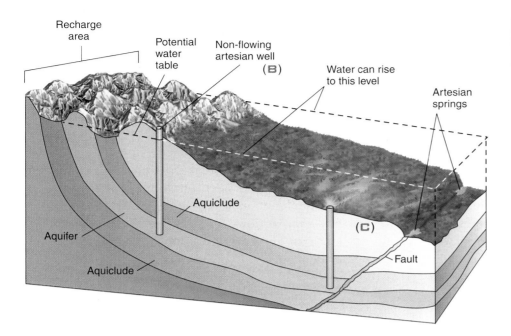

Recharge area

Potential water table

Non-flowing artesian well (B)

Water can rise to this level

Artesian springs

Aquiclude

(C)

Aquifer

Aquiclude

Fault

This diagram illustrates the natural conditions that produce artesian wells or springs. Note that the water enters the aquifer at a higher elevation than the well. As a result, the water flows or gushes out of the well with positive water pressure.

be pumped. This is called an *artesian well*. A fault can also serve as a natural conduit for artesian water.

A change in permeability of the rocks at ground level often gives rise to a spring, which is a natural analogue of an artesian well. Such a change may be due to the presence of an aquiclude (see FIGURE 5.20) or it may happen along the trace of a fault (see Figure 5.19).

What causes springs? FIGURE 5.20

A This spring in the Grand Canyon is fed by water from the porous Redwall and Muav Limestone. These cavernous limestones are the water source for many springs. The impermeable shale unit beneath them, the aquiclude, is the Bright Angel Shale.

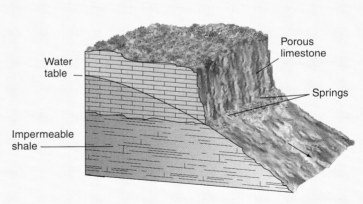

Water table

Porous limestone

Springs

Impermeable shale

B Water flows from a spring in a limestone aquifer underlain by an impermeable shale aquiclude.

GROUNDWATER DEPLETION AND CONTAMINATION

A well will supply water if it is deep enough to intersect the water table. As shown in FIGURE 5.21, a shallow well may become dry during periods when the water table is low, whereas a deeper well may yield water throughout the year. When water is pumped from a well, a *cone of depression* (a cone-shaped dip in the water table) will form around the well. In most small domestic wells, the cone of depression is hardly discernible. Wells pumped for irrigation and industrial uses, however, sometimes withdraw so much water that the cone may become very wide and steep and can lower the water levels in surrounding wells. When large cones of depression from pumped wells overlap, the result is regional depression of the water table.

If the rate of withdrawal of groundwater regularly exceeds the rate of natural recharge, the volume of stored water steadily decreases; this is called *groundwater mining*. It may take hundreds or even thousands of years for a depleted aquifer to be replenished. The results of excessive withdrawal include lowering of the water table; drying up of springs and streams; *compaction* of the aquifer; and *subsidence*, a decline in land surface elevation. Sometimes it is possible to recharge an aquifer by pumping water into it. In other cases, the effects of depletion may be permanent. When an aquifer suffers compaction—that is, when its mineral grains collapse on one another because the pore water that held them apart has been removed—it is permanently damaged and may never be able to hold as much water as it originally held.

Wells: Year-round and seasonal FIGURE 5.21

Seasonal changes affect the height of the water table. A well will produce year-round only if it extends into the year-round zone of saturation.

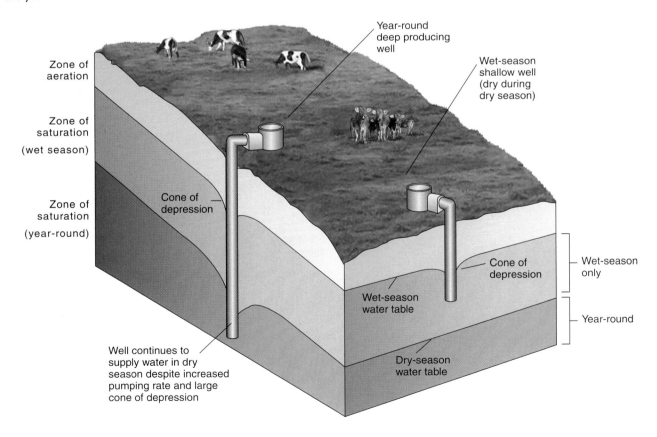

A

▲ A Newtown Creek runs for about 5 km through New York City, where it forms part of the border between Queens (at right) and Brooklyn (at left).

B The surface of the creek is constantly fouled with oil, from an underground spill that took place more than half a century ago. According to a Coast Guard study, in 1948 approximately 65 million liters of oil leaked from a refinery owned by the Standard Oil Company of New York (later renamed Mobil). (For comparison, this is more than the amount of oil spilled in the much more famous *Exxon Valdez* accident in 1989.) The successor company, ExxonMobil Company, has cleaned up part of the mess, but most of the spill continues to contaminate the groundwater and ooze to the surface in Newtown Creek. ▶

The forgotten oil spill FIGURE 5.22

B

Laws and policies relating to water rights are very complicated, and the application to groundwater is even more complicated than for surface water. Because groundwater is hidden from view, it is difficult to monitor its flow and regulate its use. If you drill a well into an aquifer underlying your property, are you entitled to withdraw as much water as you need from that well? Should you withdraw water only for your own purposes, or should you be permitted to withdraw the water and sell it elsewhere? What happens if withdrawing the groundwater depletes the aquifer and your neighbor's well runs dry?

Many of the types and sources of contaminants that affect surface water also cause groundwater contamina-tion. Because of its hidden nature, however, groundwater contamination can be much more difficult to detect, control, and clean up than surface water contamination. The most common source of water pollution in wells and springs is untreated sewage. Agricultural pesticides and fertilizers, significant sources of surface water pollution, are also common contaminants of groundwater. Harmful chemicals leaking from waste disposal facilities can also infiltrate into groundwater reservoirs and contaminate them. Probably the most serious groundwater contamination problem in North America is caused by leaking underground storage tanks at gas stations, refineries, and other industrial settings (see FIGURE 5.22).

WHEN GROUNDWATER DISSOLVES ROCKS

As discussed in Chapter 4, all rainwater and most groundwater is slightly acid due to solution of carbon dioxide. In regions underlain by rocks that are highly susceptible to chemical weathering by acid waters, groundwater creates extensive systems of underground caverns. In such areas, a distinctive landscape forms on the surface. *Karst topography*, named for the Karst region of the former Yugoslavia (now Slovenia), is characterized by many *sinkholes* (small, closed basins) and disrupted drainage patterns (see **FIGURES 5.23** and **5.24**). Streams disappear into the ground and join the groundwater. Large springs form where the water table intersects the land surface, or along favorable routes of escape, such as faults. Karst is most typical of regions underlain by soluble carbonate rocks (limestone and dolostone), although it can also occur in regions with extensive evaporite (salt) deposits. In carbonate terrains with karst topography, the rate of dissolution is faster than the average rate of erosion of surface materials by streams and mass wasting.

Caves and sinkholes

Caves and **caverns** are formed when circulating groundwater at or below the water table dissolves carbonate rock. The process begins with dissolution along interconnected fractures and bedding planes. A cave passage then develops along the most favorable flow route.

> ■ **cave** and **cavern**
> Underground open space; a cavern is a system of connected caves.

The development of a continuous passage by slowly moving groundwater may take up to 10,000 years, and the further enlargement of the passage by more rapidly flowing groundwater needed to create a fully developed cave system may take an additional 10,000 to 1 million years. Finally, the cave may become accessible to humans after the water table drops below the floor of at least some of the chambers.

In the parts of the cave that lie above the water table (in the vadose zone), groundwater continues to percolate downward, dripping from the ceiling to the floor. Calcium carbonate dissolved in the water precipitates out of solution and builds up beautiful icicle-like decorations on the cave walls, ceilings, and floor. These include stalactites (hanging from the ceiling) and sta-

Evolution of a karst landscape FIGURE 5.23

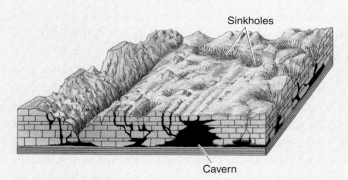

A Over time, rainwater dissolves limestone, producing caverns and sinkholes. In warm, humid climates, solution of pure limestone can form towers (left side of diagrams).

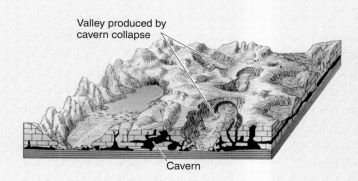

B Eventually, the caverns collapse, leaving open, flat-floored valleys. Surface streams flow on shale beds beneath the limestone. Some parts of the flat-floored valleys can be cultivated.

A These sinkholes in Florida are part of a karst terrain. Most of Missouri and large parts of Kentucky, Tennessee, Texas, and New Mexico also have carbonate karst landscapes. In Arecibo, Puerto Rico, the world's largest radio telescope was built in a sinkhole, taking advantage of its naturally circular shape.

B This is a karst terrain near Guilin, China. White limestone is visible in the pillars that remain after cavern collapse created the valley in which the village is located. The pillars are riddled with caves and passageways.

Karst FIGURE 5.24

lagmites (projecting upward from the floor), as well as columns, draperies, and flowstones. In the saturated zone, below the water table, water movement is largely horizontal, creating tubular passages.

Caves are dissolution cavities that are closed to the surface or have only a small opening. In contrast, a *sinkhole* is a dissolution cavity that is open to the sky. Some sinkholes are formed very abruptly when the roofs of caves collapse. Most sinkholes, though, develop much more slowly and less catastrophically, simply growing wider over time as the carbonate bedrock slowly dissolves.

CONCEPT CHECK STOP

What is the difference between an unconfined and a confined aquifer?

What is a water table? What lies above and below the water table?

What are some of the problems associated with the use of groundwater resources?

How and why do caves and sinkholes develop in certain regions?

Amazing Places: Lechuguilla Cave

In 1986, cavers (also known as *spelunkers*) discovered the deepest known cave in the United States, called Lechuguilla Cave, in Carlsbad Caverns National Park. Its entrance had been known for decades, but it had been considered a dead end until cavers dug through the floor to a huge network of passages on the other side.

Lechuguilla Cave has now been explored to a depth of 475 m, and has almost 160 km of mapped passages. It is as spectacular as it is deep, but is closed to the public to preserve its unusual formations.

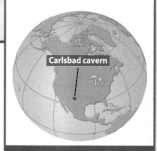

Global Locator

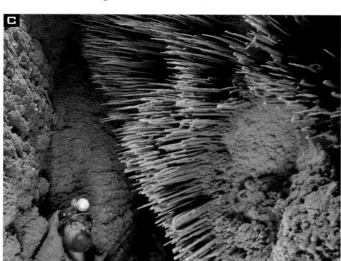

◀ A A "bush" made of fragile aragonite pokes out of a stalagmite made of calcite.

B "Soda straws" reach down from the ceiling in this small chamber. Water flows down through the center of the straw. If water starts flowing down the outside, it will build a stalactite. ▶

◀ C The origin of this rare formation, called "pool fingers," is a mystery, perhaps related to bacterial activity. The "fingers" crystallized in a pool of water and were left behind when the water retreated.

D Gypsum crystals provide a clue to this cave's unusual ▶ history. Unlike most limestone caves, which are formed by carbonic acid in rainwater, Lechuguilla formed from the bottom up. Hydrogen sulfide from lower-lying oil deposits percolated up into the groundwater and formed sulfuric acid, which dissolved away the rock. As the water table dropped, gypsum deposits precipitated out of the acidic water.

NATIONAL GEOGRAPHIC

SUMMARY

1 The Hydrologic Cycle

1. The **water cycle**, or **hydrologic cycle**, describes the movement of water from one reservoir to another in the hydrosphere. The ocean is the largest reservoir, followed by the polar ice sheets. The largest reservoir of unfrozen fresh water is groundwater. Surface water bodies, the atmosphere, and the biosphere are much smaller water reservoirs.

2. The pathways or processes by which water moves from one reservoir to another include **evaporation**, **condensation**, **precipitation**, **transpiration**, **infiltration**, and **surface runoff**.

3. Like the tectonic and rock cycles, the hydrologic cycle is a *closed cycle* of *open systems*. Because it is closed, the global hydrologic cycle maintains a mass balance.

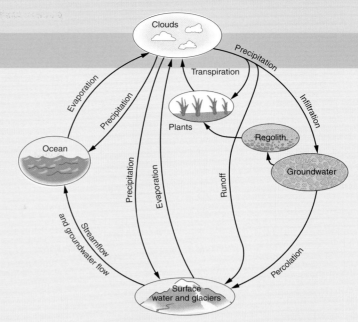

HYDROLOGIC CYCLE

2 How Water Affects Land

1. **Streams** and **rivers** flow downslope along a clearly defined natural passageway, the **channel**. Interrelated factors that influence the behavior of a stream include the **gradient**, **discharge**, **load**, and velocity of the water.

2. Streams create landforms through *erosion* and *deposition*. Straight, braided, and *meandering* channels and *oxbow lakes* are erosional landforms. *Point bars, alluvial fans, deltas,* and **floodplains** are depositional landforms made

of recently deposited sediment, or **alluvium**.

3. Every stream is surrounded by its **drainage basin**, the total area from which water flows into the stream. The

topographic "high" that separates adjacent drainage basins is a divide. Streams originate in a *headwater* region, where they collect water from a large number of smaller *tributary streams*. Water exits a stream system from the stream's *mouth,* which may empty into either a larger river, a lake, or an ocean.

4. *Lakes* form in topographic basins created by faults, glacial debris, landslides, and lava flows, as well as in areas of poor drainage. Lakes are usually short-lived features, because they tend to disappear by erosion of the outlet or by silt deposition.

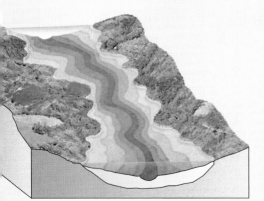

3 Surface Water as a Hazard and a Resource

1. A **flood** occurs when a stream's discharge becomes so great that it exceeds the capacity of the channel, causing the stream to overflow its banks. The risk of floods is increased by *subsidence,* and flooding can be exacerbated by human activity.

2. Prediction of flooding is based on analysis of the frequency of occurrence of past events and on real-time monitoring of storms using a hydrograph. River

channel modifications made for purposes of flood control and protection, as well as for navigation and other purposes, are collectively known as *channelization*.

3. Agriculture is by far the largest consumer of fresh water. Surface water can be transported from one drainage basin to another but only at the risk of long-term environmental changes in the affected basins.

4 Fresh Water Underground

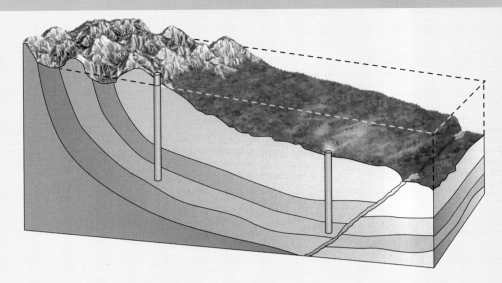

1. **Groundwater** is subsurface water contained in spaces within bedrock and regolith. In the *zone of aeration* (*unsaturated* or *vadose zone*), water is present, but does not completely saturate the ground. In the *phreatic* or *saturated zone,* all openings are filled with water. The top of the saturated zone is the **water table**.

2. The rate of groundwater flow is dependent on the characteristics of the rock or sediment through which the water must move. **Porosity** is the percentage of the total volume of a body of rock or regolith that consists of open spaces. **Permeability** is a measure of how easily a solid allows fluids to pass through it. Groundwater in the saturated zone moves slowly by **percolation** through very small pores from areas where the water table is high to where it is lower.

3. Groundwater **recharge** occurs when rainfall and snowmelt infiltrate and percolate downward to the saturated zone. **Discharge** occurs where subsurface water leaves the saturated zone and becomes surface water in a stream, lake, or **spring**.

4. An **aquifer** is a body of permeable rock or regolith in the zone of saturation. An *unconfined aquifer* is in contact with the atmosphere through the pore spaces of overlying rock or regolith, while a *confined aquifer* lies between layers of impermeable rock, called **aquicludes**. The water pressure in a confined aquifer may be high enough to force the water partway or all the way to the surface when a well is drilled into it. Such a well is called *artesian*. Excessive withdrawal from an aquifer can cause the water table to drop, springs and streams to dry up, and regolith to compact and subside.

5. Many of the types and sources of contaminants that affect surface water also cause groundwater contamination. These include untreated sewage, agricultural pesticides and fertilizers, and leaks or spills of chemicals from commercial and industrial sites.

6. Caves, caverns, sinkholes, and karst topography are formed when rocks—most commonly carbonate rocks—are dissolved by circulating groundwater, creating underground cavities. Stalactites and stalagmites are built up from calcium carbonate precipitated from percolating groundwater.

KEY TERMS

CRITICAL AND CREATIVE THINKING QUESTIONS

1. List as many ways as you can in which we depend on the availability of fresh water in our daily lives.

2. Investigate the ways in which water shortages, floods, or other processes associated with water (such as erosion and deposition of sediment) have affected human history. How did societies respond, and what effects did those responses have?

3. The concept of residence time is very important in Earth science. It applies not only to water but also to any substance that moves from one reservoir to another in the Earth system. What other substances, besides water, move around in the Earth system? Why would we want to monitor these substances and keep track of their residence times?

4. What evidence has been gathered for water-shaped landscapes on Mars or elsewhere in the solar system? How are these landforms similar to or different from their analogues on Earth?

5. Where does your community obtain its water supply? Is it a groundwater or a surface fresh-water source? Is either the quantity or the quality of the water threatened?

6. Visit a stream before and after an intense rainfall. Observe the gradient, discharge, load, and velocity of the streamflow. What changes do you notice?

What is happening in this picture ?

This 100-m-wide sinkhole opened up one day in Winter Park, Florida, and grew to the point where it eventually swallowed up part of a house, six commercial buildings, and the municipal swimming pool.

- What could have caused this to happen?

- What kind of rock is prone to this sort of collapse?

1. On this illustration, label each stage of the hydrologic cycle (1 through 6) using the following terms:

precipitation cloud formation through condensation

surface runoff surface evaporation and transpiration

infiltration evaporation from the ocean

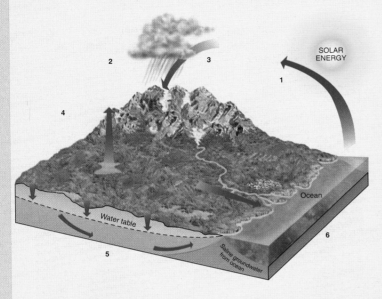

2. The illustration above depicts Earth's hydrologic cycle. Which reservoir holds most of the fresh-water resources of the planet?

a. oceans
b. ice sheets
c. lakes
d. groundwater

3. The _____ of a stream will have a large influence over channel development and the evolution of associated landforms.

a. discharge
b. gradient
c. sediment load
d. All of the above statements are correct.

4. _____ will form in streams when there is a low slope gradient and large and variable sediment load.

a. Straight channels
b. Meandering channels
c. Braided channels

5. Lakes are ephemeral features that disappear by _____.

a. the accumulation of inorganic sediment carried in by streams
b. the accumulation of organic matter produced by plants within the lake
c. gradually being drained by stream outlets eroding to lower levels
d. All of the above statements are correct.

6. Oxbow lakes form when _____.

a. levees collapse and cause floods
b. meanders form cut-offs
c. meanders fill with sediment
d. point bars collapse

7. A(n) _____ is the total area from which water flows into a stream.

a. alluvial fan
b. braided channel
c. water cycle
d. drainage basin

8. Urban development can lead to increased risk of flooding by _____.

a. compressing underlying sediments causing subsidence
b. increasing surface runoff
c. channeling runoff more quickly to rivers through storm drains
d. All of the above are correct.

9. There is a _____ chance of a 50-year flood occurring in any given year.

a. 1%
b. 2%
c. 5%
d. 10%
e. 50%

10. Though the proportions vary from one region to another, globally what accounts for the greatest demand for water?

  a. industry
  c. crop irrigation
  b. domestic use

11. What factors control the porosity of an aquifer?

  a. size of the particles in the material
  b. shape and uniformity of the material
  c. amount of cementing agent in the pore space of the material
  d. All of the above are correct.

12. A(n) _____ is an underground reservoir of water that is overlain by impermeable rock units.

  a. aquiclude
  c. confined aquifer
  b. aquifer
  d. unconfined aquifer

13. In a well drilled into a(n) _____, the water will rise to the level of the surrounding water table.

  a. aquiclude
  c. confined aquifer
  b. aquifer
  d. unconfined aquifer

14. Placing a well in a position where Earth's surface is below the water table will result in a well that _____.

  a. requires pumping to produce water
  b. does not require pumping to produce water

15. On this illustration, label each groundwater feature using the following terms:

  dry-season water table   additional wet season saturation
  wet-season water table   year-round saturation
  zone of saturation

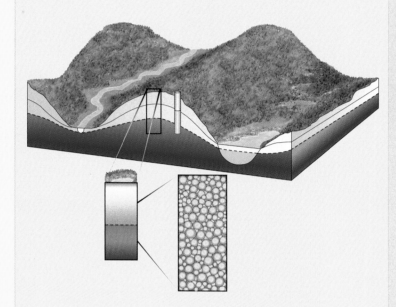

Extreme Climatic Regions: Deserts, Glaciers, and Ice Sheets

6

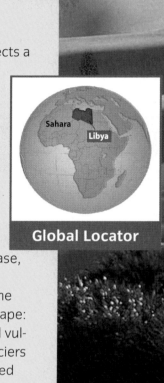

Global Locator

The Mandara Lakes in Libya excite the senses. Flat as a mirror, the water reflects a turquoise sky. The palm trees are dwarfed by the dunes of the Sahara Desert, which creep up to within a few feet of them. It is easy to see why explorers have always loved the desert: Its vastness and the beauty of its oases make human creations seem trivial by comparison.

Explorers of another stripe, like the mountaineer exulting at the summit of Aurora Peak in Alaska (inset), have always loved the world's icy wildernesses—in this case, the mountaineer is looking directly at Black Rapids Glacier. Almost every description of the Sahara Desert could also apply to this landscape: abstract, beautiful, immense, remote . . . and vulnerable. For Earth scientists, deserts and glaciers are linked by more than beauty; they are joined by extreme climates. Erosional forces such as ice and wind, which play lesser roles elsewhere, are dominant in these regions.

Deserts and glaciers are important as sensitive indicators of climate change. Glaciers and ice caps are retreating, a trend that many climatologists believe will continue throughout the 21st century. It is a trend that many people think humans may be partially responsible for because of our production of "greenhouse gases." Likewise, deserts grow or shift, in part because of changes in rainfall, and in part because of the way humans manage soil and water resources. Every one of us has good reason to pay attention to deserts and glaciers, even if we never set foot in the Sahara sand or on an Alaskan mountaintop.

The location is the Rift Valley in Kenya, in Africa, and the view is to the north. What we see is a beautiful illustration of rifting due to continental crust being spilt apart by a spreading center. The steep slopes to the right appear to be surface expressions of faults that mark the edge of the Somali plate.

- Are they normal faults or reverse faults?

Land to the left of the scarps has dropped down, forming the flat-bottomed farming valley. Still farther to the left, and out of view, is the eastern edge of the African plate. The rift valley ranges in width from 30 to 60 km.

- What will happen if spreading continues and the valley widens?

An Earth scientist would also make note of the way local agriculture exploits the stepped landscape at the edge of the Rift Valley with cultivated fields laid out in the flat tops of terraces.

The flat faces of the slopes on the fault scarps between gullies are likely to be remnants of the original fault planes.

1. The work of Earth scientists over the years has supported Wegener's contention that the current continental masses were assembled into a single supercontinent, which Wegener called _____.

   a. Pangaea
   b. Transantarctica
   c. Gondwana
   d. Tethys
   e. Laurasia

2. Which of the following lines of evidence supporting continental drift was not used by Wegener when he first proposed his hypothesis?

   a. The apparent fit of the continental margins of Africa and South America
   b. Ancient glacial deposits of the Southern Hemisphere
   c. The apparent polar wandering of the north magnetic pole
   d. Close match of ancient geology between West Africa and Brazil
   e. Close match of ancient fossils on continents separated by ocean basins

3. Analysis of apparent polar wandering paths led geophysicists to conclude _____.

   a. that Earth's magnetic poles have wandered all over the globe in the last several hundred million years
   b. that the continents had moved because it is known that the magnetic poles themselves are essentially fixed
   c. that the apparent wandering path of a continent provides a historical record of the position of that continent over time
   d. Both b and c are correct.

4. _____ is the process by which oceanic crust splits and moves apart along a midocean ridge and new oceanic crust forms.

   a. Continental drift
   b. Paleomagnetism
   c. Seafloor spreading
   d. Continental rifting

5. This illustration is a map showing the age of the seafloor, across the northern extent of the Atlantic Ocean. The Mid-Atlantic Ridge can be seen stretching roughly north–south (in the yellow band) down the middle of the map. Yellow through red colors show rocks of similar age; number them from 1 (oldest) through 5 (youngest).

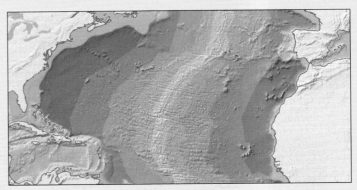

6. _____ technology has allowed scientists to measure the movement of continental crust.
   a. Global Positioning System (GPS)
   b. Seismic recording
   c. Magnetometer
   d. Gravity meter

7. This map shows radiometric ages for the Hawaiian island chain in the middle of the Pacific plate. These islands formed over a hot spot in Earth's mantle. Draw an arrow on the map showing the direction of movement of the Pacific plate over this hot spot as indicated by the ages of the islands in the Hawaiian chain.

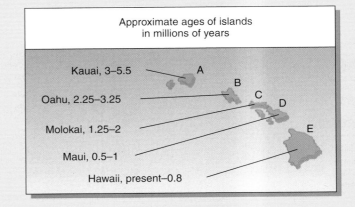

Approximate ages of islands in millions of years

Kauai, 3–5.5   A
                B
Oahu, 2.25–3.25   C
                    D
Molokai, 1.25–2
                    E
Maui, 0.5–1
Hawaii, present–0.8

8. At a _____, two lithospheric plates slide past one another horizontally.

a. divergent boundary
b. transform fault boundary
c. subduction zone boundary
d. continental collision boundary

9. At a _____, oceanic crust is consumed back into the asthenosphere.

a. divergent boundary
b. transform fault boundary
c. subduction zone boundary
d. continental collision boundary

10. At a _____, new oceanic crust forms along midocean ridges.

a. divergent boundary
b. transform fault boundary
c. subduction zone boundary
d. continental collision boundary

11. A _____ is a convergent margin along which subduction is no longer active and high mountain ranges are formed.

a. divergent boundary
b. transform fault boundary
c. subduction zone boundary
d. continental collision boundary

12. Heat from the solid mantle is released through a process of _____.

a. polar wandering
b. paleomagnetism
c. convection
d. magnetic reversal

13. This illustration shows different types of plate boundaries. For each block diagram, choose the appropriate label from the following list:

Divergent boundary   Continental collision boundary
Transform fault boundary   Subduction zone boundary

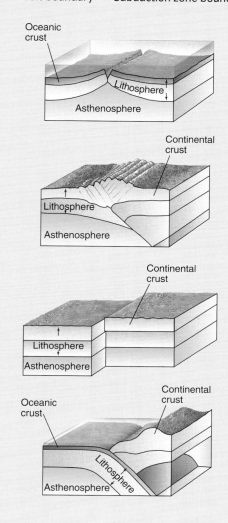

14. On the block diagrams in question 13, indicate the locations of earthquakes for each type of plate boundary. Use a red dot to show shallow-focus earthquakes and a blue dot to show the location of deep-focus earthquakes.

15. A Wilson Cycle is _____.

a. eruption and eventual subduction of basaltic lava on the seafloor
b. formation of mountains by plate collisions and subsequent erosion of the mountains
c. formation and subsequent breakup of supercontinents
d. the cycle of water through the hydrosphere

Earthquakes and Earth's Interior

On December 26, 2004, a powerful earthquake shook the Indian Ocean floor about 160 kilometers off the island of Sumatra. This earthquake—the Sumatra-Andaman earthquake of 2004—caused the largest and deadliest tsunami in history.

The quake began when part of the Indian plate, which is slipping under the Eurasian plate, suddenly slipped downward about 15 meters. The motion pushed the seafloor up as much as 5 m on the Eurasian side. Unlike most earthquakes, which are over in seconds, the slippage continued for 10 minutes as the interface between the plates slipped, section by section, for 1200 km to the north. On the surface of the ocean, waves swept toward Indonesia, Thailand, Sri Lanka, and India. As they approached the shore and the ocean shallowed, the waves grew in height; when they reached the shore, they had grown to heights of 20 or 30 m. Sweeping inland, the waves obliterated everything in their path. The photograph here shows destruction in Sumatra, and the satellite views (inset) show houses, roads, and bridges in Banda Aceh, on the island of Sumatra, before and after the tsunami.

Although earthquakes and tsunamis are common in the Indian Ocean, no warning system was in place when this disaster occurred. The resulting devastation caused hundreds of billions of dollars in damages and at least 227,000 deaths.

Earthquakes give Earth scientists a window into Earth's internal structure. We cannot prevent earthquakes, but scientists are getting better at understanding and preparing for their after effects, including devastating tsunamis. The knowledge we gain from such events allows us to prepare more effectively for future disasters.

Global Locator

BEFORE

AFTER

NATIONAL GEOGRAPHIC

HIGH-RISES RATTLE
Designed to sway, skyscrapers generally don't collapse but do sustain damage. Windows explode, girder welds crack, fires erupt.

BUILDINGS SINK
When soil is saturated, quakes can turn solid ground into a molasses-like mix that causes buildings to lean or even topple.

5 REINFORCED COLUMNS

4 PROTECTED UTILITY LINES

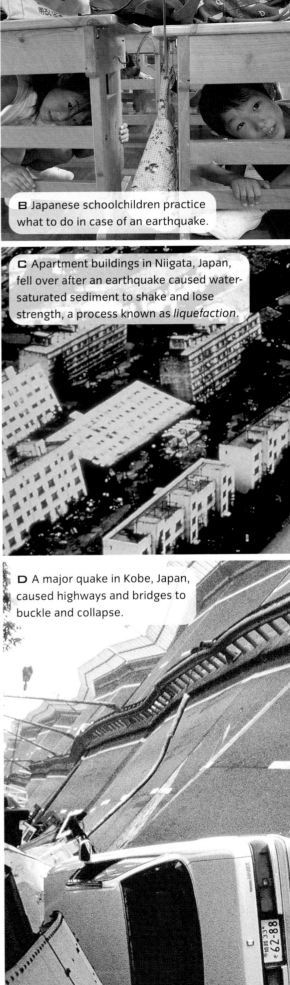

B Japanese schoolchildren practice what to do in case of an earthquake.

C Apartment buildings in Niigata, Japan, fell over after an earthquake caused water-saturated sediment to shake and lose strength, a process known as *liquefaction*.

D A major quake in Kobe, Japan, caused highways and bridges to buckle and collapse.

The Science of Seismology

S eismologists can quickly locate an earthquake anywhere on Earth and tell how strong it is. They are also very good at telling the difference between earthquakes caused by natural movements and other seismic disturbances, such as explosions and landslides. The device used to study all aspects of earthquakes is a *seismograph*.

SEISMOGRAPHS

The earliest known **seismographs** (also called *seismometers*) were invented in China in the second century (see **FIGURE 8.9A**). The first seismographs in Europe were invented much later, in the 19th century. Modern seismographs provide a printed or digital record of seismic waves, called a **seismogram** (see **FIGURE 8.9B**).

The most advanced seismographs measure the ground's motion optically and amplify the signal electronically. Vibrations as tiny as one hundred-millionth (10^{-8}) of a centimeter can be detected. Indeed, many instruments are so sensitive they can sense vibrations caused by a moving automobile many blocks away.

seismograph An instrument that detects, measures, and records vibrations of Earth's surface.

seismogram The record made by a seismograph.

Ancient and modern seismographs FIGURE 8.9

▲ A Ancient Chinese seismograph.

B Seismographs use the principle of *inertia*—the resistance of a mass to motion. In this schematic diagram, seismic waves cause the support post and the roll of paper to vibrate back and forth. However, the large mass attached to the pendulum, and the pen attached to it, barely move at all. It looks to an observer as if the pen is moving, but in reality it is the paper that moves.

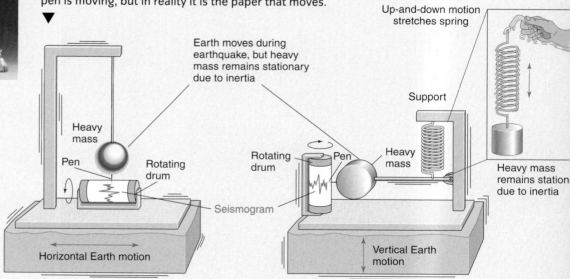

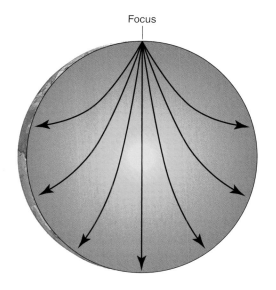

Focus

The energy released during an earthquake travels through Earth from its source (the focus). If Earth were of uniform density throughout, the waves would travel in straight lines. However, rock density increases with depth as a result of increasing pressure. Seismic waves travel faster through denser rocks; hence, they travel more quickly at greater depths. This increase in velocity with depth causes seismic wave paths to be curved, rather than straight. (This diagram is not completely accurate, because the increase in rock density and seismic velocity with depth is not smooth; you will see a more detailed diagram later in the chapter.)

SEISMIC WAVES

body wave A seismic wave that travels through Earth's interior.

surface wave A seismic wave that travels along Earth's surface.

focus The location where rupture commences and an earthquake's energy is first released.

P wave The first, or primary, wave to be detected by a seismograph.

The energy released by an earthquake is transmitted to other parts of Earth in the form of seismic waves. The waves move as elastic deformations of the rocks; they leave no record behind them once they have passed, so they must be detected while they pass. The waves, which include both **body waves** and **surface waves**, travel outward in all directions from the earthquake's **focus** (see Figure 8.10).

Body waves Following an earthquake, a seismograph will record two kinds of *body waves*; they travel at different speeds so there is a time gap between their arrivals. The first set of waves to arrive and be detected by a seismograph are called **P waves** (or primary waves). P waves are **compressional waves** (see Figure 8.11A on page 248); they are like sound waves and can pass through solids, liquids, and gases. They have the highest velocity of all seismic waves—typically 6 km/s in the uppermost portion of the crust.

The second waves to reach and be recorded by a seismograph after earthquakes are called **S waves** (or secondary waves). They travel through solid materials by an undulating, or shearing motion (see Figure 8.11B). Solids tend to resist a shear force and bounce back to their original shape afterward, whereas liquids and gases do not. Without this elastic rebound, there can be no wave. Therefore, **shear waves** cannot be transmitted through liquids or gases. This has important consequences for the interpretation of seismic waves, as you will soon see. Shear waves travel only 3.5 km/s—not as fast as compression waves.

compressional wave A seismic body wave consisting of alternate pulses of compression and expansion in the direction of travel; P wave or primary wave.

S wave The second kind of body wave to be detected by a seismograph.

shear wave A seismic body wave in which rock is subjected to side-to-side or up-and-down forces, perpendicular to the direction of travel; S wave or secondary wave.

Surface waves In addition to body waves that travel through Earth, earthquakes generate surface waves that travel along or near Earth's surface, like waves along the surface of the ocean. They travel more slowly than P and S waves, and they pass around Earth rather than through it. Thus, surface waves are the last to be

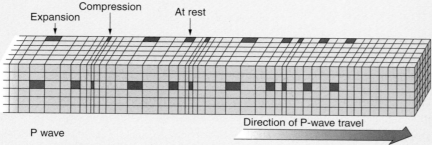

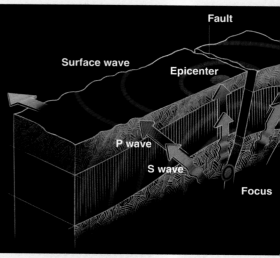

A A compressional wave alternately squeezes and stretches the rock as it passes through. The grid is intended to help you visualize how the rock responds. All the divisions in the grid start out square, but the wave alternately squeezes them down to narrow rectangles and then stretches them to long rectangles. Sound waves travel through air by the same means—alternate compressions and expansions of air.

C P and S waves travel outward from the focus, generating waves that travel along the surface.

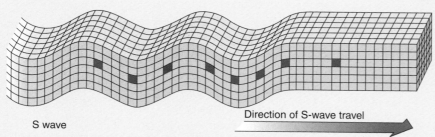

B A shear wave causes the rock to vibrate up and down, like a rope whose end is being shaken. In this case, the squares do not expand or contract but do get distorted, changing shape alternately from a square to a parallelogram and back to a square again.

Seismic body waves FIGURE 8.11

detected by a seismograph. FIGURE 8.12A shows a typical seismogram, in which the P waves' arrival is first, followed by the arrival of the S waves, and finally by that of the surface waves. There are two kinds of surface waves, and both are named for the English mathematicians who first predicted their existence. Rayleigh waves, named for Lord Rayleigh, cause Earth's surface to move up and down, like a wave on the ocean. Love waves, named for A. E. H. Love, cause the surface to shake in a sideways motion, but do not cause any vertical motion. Surface waves are responsible for much ground shaking and structural damage during major earthquakes.

LOCATING EARTHQUAKES

The **epicenter**, or surface location, of an earthquake can be determined through simple calculations using travel-time curves from at least three seismographs that have

epicenter The point on Earth's surface that is directly above an earthquake's focus.

recorded the quake. The first step is to find out how far each seismograph is from the source of the earthquake. The greater the distance traveled by the seismic waves, the more the S waves will lag behind the P waves. Thus, the lag time between the P and S waves on a seismogram (see FIGURE 8.12B) provides seismologists with the necessary distance information.

After determining the distance from each seismograph to the source of the earthquake, the seismologist draws a circle on a map, with the seismic station at the center of the circle. The radius of the circle is the distance from the seismograph to the focus. It is a circle because the seismologist knows only the distance, not the direction. When this information is calculated and plotted for three or more seismographs, the unique point on the map where the three circles intersect is the location of the epicenter (see FIGURE 8.12C). This process is called *triangulation*.

A Seismogram of a typical earthquake.

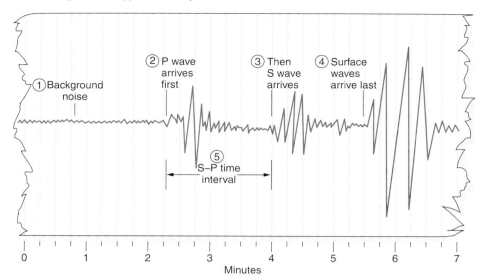

① The earthquake happens at time 0.

② The first P waves arrive a little over 2 minutes later.

③ The first S waves arrive 4 minutes later.

④ The surface waves, which travel the long way around Earth's surface, arrive last.

⑤ The S–P interval, here slightly less than 2 minutes, tells the seismologist how far away the earthquake was.

▼ **B** P and S waves leave the focus of an earthquake at the same instant. The fast-moving P waves reach a seismograph first, and sometime later the slower-moving S waves arrive. The delay in arrival time increases with distance traveled. Average travel-time curves are used to locate an epicenter. For example, when seismologists at a station measure the S–P time interval to be 13.7 min – 7.4 min = 6.3 min, they know the epicenter is 4000 km away from the station.

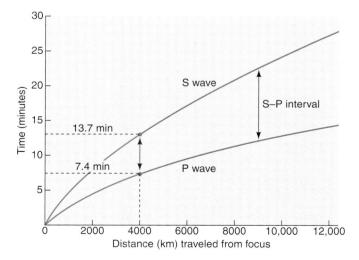

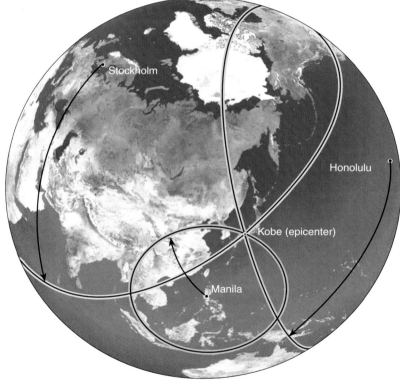

▲ **C** The method of triangulation. If three seismic stations—shown here in Stockholm, Honolulu, and Manila—record an earthquake, each one can independently determine its own distance from the focus of the quake, thus generating a circle on which the epicenter of the earthquake must lie. The three circles have a unique intersection point, which is the location of the epicenter: in this case Kobe, Japan (the site of a major earthquake in 1995). The black arrows indicate the radius of the circle on which each station estimated the epicenter must lie.

MEASURING EARTHQUAKES

Earth scientists use several different scales to quantify the strength or *magnitude* of an earthquake, by which we mean the amount of energy released during the quake. The earliest scale, developed by an Italian scientist in 1902 and later modified, is called the *Modified Mercalli Intensity Scale*. The scale is based on descriptions of vibrations that people felt, saw, and heard and on the extent of damage to buildings. The scale ranges from I (not felt, except under favorable circumstances) to XII (waves visible, practically all buildings destroyed). The Mercalli intensity of an earthquake varies with distance from the epicenter—an earthquake could have an intensity of X near the epicenter whereas a hundred kilometers away the intensity could be only II. The

> **Richter magnitude scale**
> A scale of earthquake intensity based on the heights, or amplitudes, of the seismic waves recorded on a seismograph.

Modified Mercalli Scale is most useful for studying the earthquakes that happened before the development of modern seismic equipment.

The most familiar of the modern intensity scales is the **Richter magnitude scale**. It has been superseded for most purposes by the *moment magnitude*, which uses the same scale but is computed in a somewhat different way.

The Richter magnitude scale

Charles Richter developed his famous magnitude scale in 1935. Though it was not the first earthquake intensity scale, it was an important advance because it was the first to use data from seismographs rather than subjective estimates of damage. Also, it compensated for the distance between the seismograph and the focus. This means that each seismic station will (in principle) calculate the same magnitude for a given earthquake, no matter how far from the epicenter it may be located. As discussed above, a scale such as the Modified Mercalli Scale, based on estimates of the damages sustained, yields the highest magnitude closest to the epicenter, where the damage is greatest.

The Richter scale is *logarithmic*, which means that each unit increase on the scale corresponds to a 10-fold increase in the amplitude of the wave signal. Thus, a magnitude 6 earthquake has an amplitude 10 times larger than that of a magnitude 5 quake. A magnitude 7 earthquake has an amplitude 100 times larger (10×10) than that of a magnitude 5 quake. However, even this comparison understates the difference, because the amount of damage done by an earthquake is more closely related to the amount of energy released in the quake. Each step in the Richter scale corresponds roughly to a 32-fold increase in energy (see *What an Earth Scientist Sees: Richter Magnitude: A Logarithmic Scale*). The actual damage done by a quake will also depend, of course, on local conditions, such as how densely populated the area is.

Moment magnitude

Seismologists today determine magnitudes using both the Richter scale and **moment magnitude**, which are calculated using different starting assumptions. Richter scale calculations are based on the assumption that an earthquake focus is a point. Therefore, the Richter scale is best suited for earthquakes in which energy is released from a relatively small area of a locked fault.

> **moment magnitude** A measure of earthquake strength based on the rupture size, rock properties, and amount of displacement on the fault surface.

In contrast, the calculation of seismic moment takes account of the fact that energy may be released over a large area. A classic example was the Sumatra-Andaman earthquake of 2004, when a 1200-km length of fault moved. Though the method of calculation is different, the scales are the same because they measure the same thing—the amount of energy released. In either system, magnitude 9 is catastrophic, whereas magnitude 3 is imperceptible to humans.

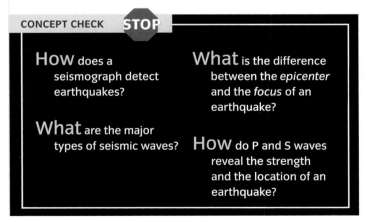

CONCEPT CHECK **STOP**

How does a seismograph detect earthquakes?

What are the major types of seismic waves?

What is the difference between the *epicenter* and the *focus* of an earthquake?

How do P and S waves reveal the strength and the location of an earthquake?

Richter Magnitude: A Logarithmic Scale

The energy released in an earthquake increases exponentially with its magnitude. A magnitude 6 earthquake releases as much energy as the atomic bomb dropped on Hiroshima, the largest ever used in combat. A magnitude 7 quake would be equivalent in energy to about 32 Hiroshima bombs, and a magnitude 8 quake would be equivalent to 32 × 32, or about 1000, of them. A magnitude 9 quake, such as the Sumatra-Andaman quake, is equivalent to 32 × 32 × 32, or about 32,000 bombs. An Earth scientist would confirm the magnitude of an earthquake determined from a seismogram by the degree of damage close to the epicenter.

Parkfield, CA, 2004

◄ **Richter magnitude 6**

Damage on surface close to the epicenter: small objects broken, sleepers awakened (Mercalli intensity ≈VII)
Energy released: about the same as one atomic bomb

Richter magnitude 7   ▶

Damage on surface close to the epicenter: some walls fall, general panic (Mercalli intensity ≈IX)
Energy released: about the same as 32 atomic bombs

San Francisco, CA, 1906

Kobe, Japan, 1995

◄ **Richter magnitude 8**

Damage on surface close to the epicenter: wide destruction, thousands dead (Mercalli intensity ≈XI)
Energy released: about the same as 1000 atomic bombs

CRITICAL THINKING

Here's an interesting question:
• What might cause there to be a limit to the largest possible magnitude for an earthquake, or is there no limit to the magnitude of an earthquake?

Looking into Earth's Interior

Although earthquakes are significant to society because of the damage they cause, they also have benefits from a scientific perspective. They provide us with some of our most detailed information about Earth's interior—including parts that we can never hope to observe directly.

When scientists cannot study something by direct sampling, a second method comes to the forefront: indirect study or *remote sensing*. Some familiar objects—including the human eye—are actually remote-sensing devices. A camera, for instance, is a remote sensing instrument that collects information about how an object reflects light. Medical techniques such as X-rays allow doctors to study the inside of the body remotely without opening it up surgically.

The seismic waves from an earthquake are much like X-rays, in the sense that they enter Earth near the surface, travel all the way through it, and emerge on the other side. They travel along different paths depending on the different kinds of materials they encounter. We will first discuss what earthquakes reveal about Earth's structure and then describe other sources of information about Earth's interior.

Process Diagram

Seismic waves in Earth's interior FIGURE 8.13

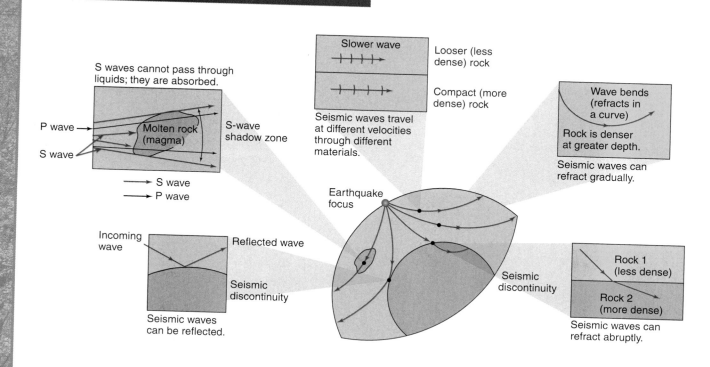

S waves cannot pass through liquids; they are absorbed.

P wave
S wave
Molten rock (magma)
S-wave shadow zone

→ S wave
→ P wave

Slower wave — Looser (less dense) rock

Compact (more dense) rock

Seismic waves travel at different velocities through different materials.

Wave bends (refracts in a curve)

Rock is denser at greater depth.

Seismic waves can refract gradually.

Earthquake focus

Incoming wave — Reflected wave

Seismic discontinuity

Seismic waves can be reflected.

Seismic discontinuity

Rock 1 (less dense)

Rock 2 (more dense)

Seismic waves can refract abruptly.

VIEW THIS IN ACTION in your WileyPLUS course

Diamonds: Messengers from the Deep

A Diamonds form at the extremely high pressures found at depths of 100 to 300 km. The different colors in this diamond—nicknamed the "Picasso" diamond—show various zones of growth. The diamond is approximately 1 millimeter across, and the colors are revealed by a special type of photography that highlights small variations in composition. An observant Earth scientist would discover that within a diamond there are tiny inclusions of other minerals, such as olivine, garnet, and graphite, that were trapped as the diamond grew. These tiny inclusions are pristine samples from deep in the mantle. ▼

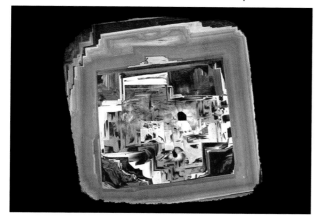

B The beautiful uncut yellow diamond is the 253.7-carat Oppenheimer Diamond from the Smithsonian Institution; it was discovered in South Africa in 1964. ▼

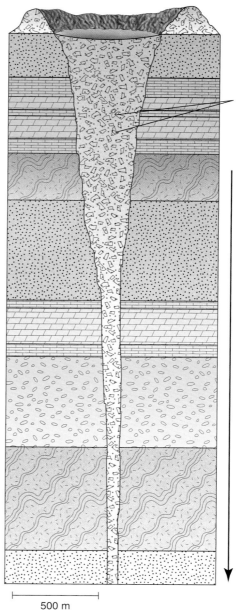

Magma vent is circular when viewed from above.

Xenoliths of mantle rock

Pipe extends 150–200 km down into mantle

|— 500 m —|

C To reach the surface from such great depths, diamonds must be carried by an eruption of unusual ferocity. These eruptions leave behind a long, cone-shaped tube of solidified magma, called a *kimberlite pipe*. Although most people treasure diamonds for their beauty and luster, Earth scientists treasure them also as "messengers"—samples from an otherwise inaccessible region of Earth's interior.

CRITICAL THINKING ?

Here's an interesting question:
- How might you distinguish a natural diamond from a synthetic one grown in the laboratory?

A Earth is surrounded by a magnetic field, which causes a compass needle to point north. More precisely, the needle is aligned along the field lines that lead to the north and south magnetic poles, which are almost—but not exactly—aligned with Earth's north (N) and south (S) geographic poles.

B This is a photograph of the aurora borealis, or northern lights, as seen from Fairbanks, Alaska. This phenomenon is caused by charged particles from the Sun entering Earth's atmosphere at high latitudes along magnetic field lines.

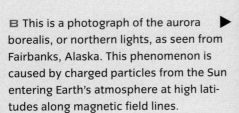

Earth's magnetic field FIGURE 8.16

Indirect observation: methods from physics, astronomy, and chemistry

The availability of indirect or remote techniques for the study of Earth's interior has increased considerably since the dawn of the Space Age. Many of these techniques are also used to study other planets in the solar system. We have already discussed the single most important method for the study of Earth's interior: the study of seismic waves generated by earthquakes. Another important geophysical method relies on the study of Earth's magnetic field.

Magnetism is a force created either by permanent magnets (ferromagnets) or by moving electrical charges. We can try thinking of Earth as having a huge dipole bar magnet with north and south poles at its center, offset slightly from the geographic North and South Poles. The problem with this analogy is that solids, including bar magnets, lose their magnetism at temperatures above a critical transition temperature, called the *Curie point*, which is specific to each material. The Curie point for iron is about 770°C, but we know that the temperature in the core is *much* higher than this—at least 5000°C.

Since there can't be a giant bar magnet inside the planet, moving electrical charges must be responsible for generating Earth's magnetic field (see FIGURE 8.16). Physicists have shown that the movement of an electrically conducting liquid inside a planet could generate a self-sustaining magnetic field, much like a rotating coil of wire in an electric motor. This is consistent with the observation from seismology that at least the outer part of Earth's core is liquid. However, molten rock is not a good enough electrical conductor to generate a magnetic field in this manner; for this and other reasons, Earth scientists believe that the liquid outer core is made of molten iron and nickel. This is consistent with evidence from meteorites, discussed later.

We can also gain a certain amount of information about any planet's interior—including Earth's—from astronomical observations. The first step is to determine the planet's *mass*. This can be deduced from the planet's gravitational influence on other planets and satellites. Second, we need to know the *diameter* of the planet. Knowing the dimensions of the planet and its shape (in the case of Earth, a very slightly flattened sphere), it is a simple matter to figure out its volume and average density (mass divided by volume).

What do these kinds of measurements reveal about Earth's interior? For one thing, we can determine whether material is distributed evenly throughout the planet. The rocks at Earth's surface are very light (low-density) compared to the planet as a whole. Surface rocks have an average density of about 2.8 g/cm³, whereas Earth's overall density is 5.5 g/cm³. (For comparison, water has a density of 1 g/cm³ at 4°C.) For the planet as a whole to have such a high density, with such low-density rocks at the surface, there must be a concentration of denser material somewhere inside the planet. Seismic evidence indicates that the density of the mantle is less than 4.0 g/cm³, so the missing mass must be in the core and the density of the core must be greater than 10 g/cm³, which is consistent with the densities of iron meteorites.

A third way to study Earth's interior is to analyze the building blocks that formed it. Planetary scientists have discovered that most (though not all) meteorites were formed at about the same time and in the same part of the solar system as Earth. Some of these meteorites are *primitive*—that is, they have remained unaffected by melting and other processes since the beginning of the solar system. These meteorites give scientists an idea of the overall composition of the solar system and its constituent bodies. Other meteorites—the *irons, stony-irons,* and some kinds of *stony meteorites*—show signs of melting and differentiation, and may be more representative of Earth's core and mantle. It is highly significant that a core with the composition of a typical iron meteorite (mostly iron and nickel) would bring Earth's overall density up to the observed value of 5.5 g/cm³.

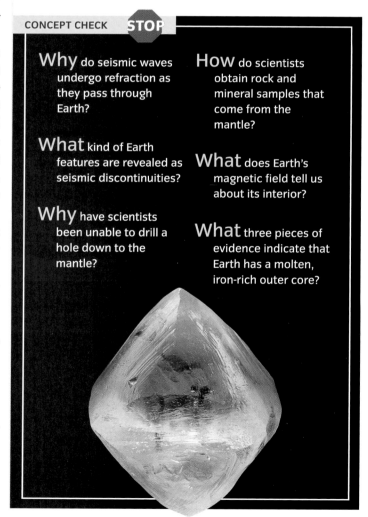

CONCEPT CHECK STOP

Why do seismic waves undergo refraction as they pass through Earth?

What kind of Earth features are revealed as seismic discontinuities?

Why have scientists been unable to drill a hole down to the mantle?

How do scientists obtain rock and mineral samples that come from the mantle?

What does Earth's magnetic field tell us about its interior?

What three pieces of evidence indicate that Earth has a molten, iron-rich outer core?

How and Why Rock Breaks

LEARNING OBJECTIVES

Explain the difference between stress and pressure.

Define the three kinds of stress.

Identify the kinds of stress associated with various types of faults.

Relate the severity of earthquakes to different types of faults.

The elastic rebound theory says that an earthquake is caused by the storage of energy in rocks that have been bent and buckled because friction stopped motion along a fault. Faults are not all alike, so it is helpful briefly to review the different kinds of faults and the earthquakes they cause.

STRESS AND STRAIN

stress The force acting on a surface, per unit area, which may be greater in certain directions than others.

pressure A particular kind of stress in which forces acting on a body are the same in all directions.

In discussing deformation and fracture of rocks, we use the word **stress** rather than **pressure**. These two words are related in meaning but different in connotation. Both are defined as the force acting on a surface per unit area. The term *pressure*, as used in Earth science, implies that the forces on a body of rock are essentially uniform in all directions. Sometimes this is called *uniform stress* or *confining stress.* These are appropriate terms to describe, for instance, the stress on a small body immersed in a liquid such as water or magma.

Rocks, however, are solids; unlike liquids and gases, solids can resist different pressures in different directions at the same time. For this reason, *stress* is a more versatile term for discussing rock deformation, because it does not imply that the forces are necessarily the same in all directions. To be even more precise, we sometimes use the term *differential stress* when the force is greater from one direction than from another. The stresses that cause rock to fracture or change shape are differential. They can be classified into three different kinds, as illustrated in **FIG-URE 8.17**: **tension**, **compression**, and **shear**.

In response to stress, a rock will experience **strain**. Uniform stress causes a change in volume only, while differential stress may cause a change in shape. For example, if a rock is subjected to uniform stress by being buried deep in Earth, its volume will decrease; that is, the rock will be compressed but it is not likely to fracture.

tension A stress that acts in a direction perpendicular to and *away* from a surface.

compression A stress that acts in a direction perpendicular and *toward* a surface.

shear A stress that acts in a direction *parallel* to a surface.

strain A change in the shape or volume of a rock in response to stress.

KINDS OF DEFORMATION

The way a rock responds to differential stress depends not only on the amount and kind of stress but also on the nature of the rock itself. For example, a rock may stretch like a metal spring and then return to its original shape when the stress is removed. Such a nonper-

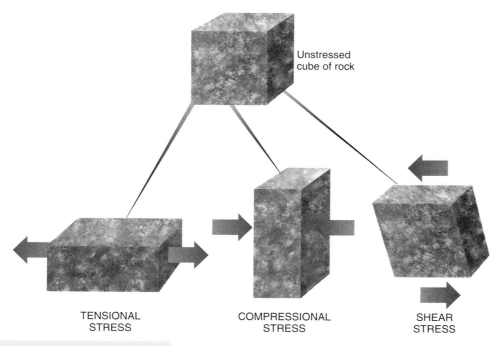

Unstressed cube of rock

TENSIONAL STRESS

COMPRESSIONAL STRESS

SHEAR STRESS

Three types of stress FIGURE 8.17

The shape of a cube of rock changes, depending on the type of stress applied to it. The arrows indicate tensional, compressional, and shear stress. Rocks that are subjected to differential stress—stress that is stronger in one direction than another—typically respond by changing their shape, as shown by these blocks.

manent change is called **elastic deformation**. For most solids, including rocks, there is a degree of stress—

> **elastic deformation** A temporary change in the shape or volume from which a material rebounds after the deforming stress is removed.

called the *elastic limit*—beyond which the material is permanently deformed. If the rock is subjected to more stress than this, it will not return to its original size and shape when the stress is removed.

When rocks are stretched past their elastic limit, they can deform in two different ways. **Ductile deformation**, also called *plastic deformation*, is one type of permanent deformation in a rock (or other solid) that has been stressed beyond its elastic limit. Alternatively, the rock may undergo **brittle deformation**. A brittle material deforms by fracturing, whereas a ductile material deforms by changing its

shape. Drop a piece of chalk on the floor and it will break. Drop a piece of play dough, and it will bend or flatten instead of breaking. Under the conditions of room temperature and atmospheric pressure, chalk is brittle, and play dough is ductile. Rock in the lithosphere is brittle, like chalk, and so deformation in the lithosphere is commonly by fracture. Rock in the asthenosphere and mesosphere is very hot and, like play dough, has little strength, so deformation in these regions is ductile. Faults and earthquakes are confined to the lithosphere.

> **ductile deformation** A permanent but gradual change in the shape or volume of a material, caused by flowing or bending.

> **brittle deformation** A permanent change in shape or volume, in which a material breaks or cracks.

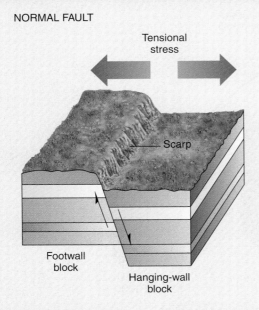

NORMAL FAULT

Tensional stress

Scarp

Footwall block

Hanging-wall block

A When the crust is stretched by tension, normal faults occur. The block on the overhanging part of the fault moves down relative to the block underneath the fault.

REVERSE FAULT

Compressional stress

Lake

Footwall block

Hanging-wall block

B In a reverse fault, compression pushes the overhanging block up and over the one underneath.

KINDS OF FAULTS

A *fault,* as defined in Chapter 7, is a fracture along which movement has occurred. Most faults are small, only meters long, but others are very large, and can be 100 km or more in length (see *Amazing Places: Loch Ness* on page 264).

Different kinds of faults are caused by different kinds of stress (see FIGURE 8.18). Tensional (or *extensional*) stress pulls the crust apart and causes **normal faults** (FIGURE 8.18A). Rocks are weak when subjected to tensional forces—that is, they deform very little before they break. As a consequence, earthquakes associated with normal faults tend to be of low magnitude and have shallow foci. Such earthquakes

> **normal fault**
> A fault in which the block above the fault surface moves down relative to the block below.

are common along spreading centers between two plates.

Rocks are strong when compressed—that is, they can store a lot of elastic energy before they break. Compressional stress is responsible for **reverse** and **thrust faults**, and for Earth's largest magnitude earthquakes (FIGURES 8.18B and 8.18C). In reverse faults, one side of the fault rides up over the other side. Thrust faults are reverse faults that have gentle slopes. Reverse faults and thrust faults are common along convergent plate boundaries, and earthquakes associated with them tend to be of large magnitude and often have deep foci.

> **reverse fault**
> A fault in which the block on top of the fault moves up and over the block on the bottom.

> **thrust fault** A reverse fault that cuts Earth's surface at a shallow angle.

- earthquake p. 234
- seismology p. 234
- elastic rebound theory p. 235
- seismic wave p. 235
- paleoseismology p. 241
- seismograph p. 246
- seismogram p. 246
- body wave p. 247
- surface wave p. 247

- focus p. 247
- P wave p. 247
- compressional wave p. 247
- S wave p. 247
- shear wave p. 247
- epicenter p. 248
- Richter magnitude scale p. 250
- moment magnitude p. 250

- seismic discontinuity p. 253
- refraction p. 254
- reflection p. 254
- stress p. 260
- pressure p. 260
- tension p. 260
- compression p. 260
- shear p. 260
- strain p. 260

- elastic deformation p. 261
- ductile deformation p. 261
- brittle deformation p. 261
- normal fault p. 262
- reverse fault p. 262
- thrust fault p. 262
- transform fault p. 263

CRITICAL AND CREATIVE THINKING QUESTIONS

1. Use the elastic rebound theory to describe what happens to rocks at the focus just before, during, and after an earthquake.

2. Why is short-term prediction of earthquakes so much less successful than long-term prediction? Why do you think seismologists are extremely cautious about making predictions? Do you think it will ever be possible to predict earthquakes accurately? Research your answer.

3. If you were asked to determine the exact shape and size of Earth, how would you go about it? What would you do differently if you were not allowed to use Space Age technology such as satellite photographs and orbital data?

4. Some of the boundaries inside Earth represent transitions between layers with differing compositions, whereas others rep-

resent transitions between layers with different physical states. Find out more about these different layers, and draw a detailed diagram to show the layering.

5. Which of the techniques used to study Earth's interior could also be used to study other planets? Which ones cannot, and why? Scientists know more about the surface of the Sun than about the interior of our own planet; why do you think this is so?

6. Find real examples of plate boundaries along which each of the following types of stress predominates: (a) compression, (b) tension, and (c) shearing. Try to find different examples than those used in the text.

What is happening in this picture ?

This photograph shows a stream in Carrizo Plains, California. The stream makes an abrupt turn to the right and then a 90-degree turn to the left.

- What reason can you suggest for this stream's strange behavior?

1. At which type of plate boundary have the largest recorded earthquakes occurred?
   a. divergent boundaries
   b. transform fault boundaries
   c. subduction zone boundaries
   d. continental collision boundaries

2. According to the elastic rebound model, earthquakes are caused by _____.
   a. the slow release of gases from the athenosphere
   b. the sudden release of energy stored in rocks through continuing stress
   c. the sudden movement of otherwise stable tectonic plates
   d. the rapid release of gases from the asthenosphere

3. _____ and the resulting collapse of buildings, bridges, and other structures are usually the most significant primary hazards to cause damage during an earthquake.
   a. Fire
   b. Tsunami
   c. Ground liquefaction
   d. Ground shaking

4. _____ can provide a good idea of which regions are at risk for severe earthquakes.
   a. Short-term forecasting
   b. Long-term forecasting
   c. Unusual animal behavior studies
   d. Studies of groundwater levels

5. Body waves _____.
   a. move through Earth's interior
   b. cannot penetrate Earth's liquid outer core
   c. move along Earth's surface, causing great destruction
   d. Both b and c are correct.

6. Illustrations **A** and **B** depict two different types of seismic waves. Which of the following statements can be made about these two seismic waves?
   a. The wave depicted in A is a P wave and has a greater velocity through Earth's crust than other types of seismic waves.
   b. The wave depicted in A is an S wave and has a greater velocity through Earth's crust than other types of seismic waves.
   c. The wave depicted in B is a P wave and has a greater velocity through Earth's crust than other types of seismic waves.
   d. The wave depicted in B is an S wave and has a greater velocity through Earth's crust than other types of seismic waves.

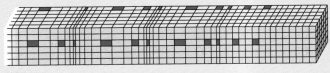

A

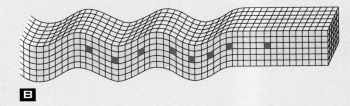

B

7. This illustration shows a seismogram of a hypothetical earthquake. On the seismogram, label the following:

   S–P interval   First arrival of P wave
   First arrival of S wave   Background noise
   First arrival of surface waves

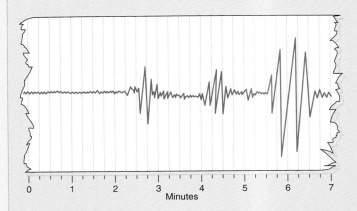

8. Using seismograms from three different seismic recording stations, A, B, and C, you determine the epicenter of an earthquake. Stations A and B both had an S–P interval of 3 seconds while C had an S–P interval of 11 seconds. Which of the following statements most accurately depicts the location of the epicenter?
   a. The epicenter is closest to Station A and equally far from B and C.
   b. The epicenter is closest to Station B and equally far from A and C.
   c. The epicenter is closest to Station C and equally far from A and B.
   d. The epicenter is equally close to A and B and farthest from Station C.

9. For the earthquake mentioned above, which seismic recording station would have recorded the P wave first?
   a. Station A
   b. Station B
   c. Station C
   d. Both stations A and B would have recorded the P wave before station C.

10. A magnitude 8 earthquake releases approximately _____ times more energy than a magnitude 7 event.
   a. 2   c. 20   e. 31.5
   b. 10   d. 21.5

11. Moment magnitude differs from Richter magnitude _____.
   a. because the Richter magnitude assumes earthquakes are generated at a point source, whereas moment magnitude takes into account that earthquakes can be generated over a large area of rupture
   b. because the moment magnitude assumes earthquakes are generated at a point source, whereas Richter magnitude takes into account that earthquakes can be generated over a large area of rupture
   c. in that moment magnitude uses Roman numerals to designate strength of an earthquake
   d. in that Richter magnitude uses Roman numerals to designate strength of an earthquake

12. When seismic waves reach a discontinuity inside Earth's interior, _____.
   a. they can be refracted, or bent, as they pass from the first material into the second
   b. they can be reflected, which means that all or part of the wave energy bounces back
   c. they can be absorbed, which means that all or part of the wave energy is blocked by the second material
   d. All of the above statements are correct.

13. In _____, rocks will bend as long as stress is applied to the crust, but resume their original shape if the stress is released.
   a. elastic deformation
   b. brittle deformation
   c. ductile deformation

14. This diagram shows a faulted block of Earth's crust. What type of fault is depicted in the diagram?
   a. normal fault
   b. thrust or reverse fault
   c. strike-slip fault

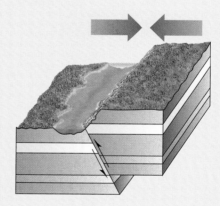

15. The fault depicted in the diagram in question 14 must have formed in response to what kind of stress?
   a. tension
   b. compression
   c. shear

Volcanism and Other Igneous Processes

9

The eruption of Mt. Pinatubo, on the island of Luzon in the Philippines, on June 15, 1991, was the second-largest volcanic eruption of the 20th century and the largest in a densely populated area. The top of the mountain exploded (main photo), blasting a gaping hole 2.5 kilometers in diameter and propelling volcanic ash and sulfurous gases more than 30 km into the atmosphere. The cloud lingered in the stratosphere and lowered worldwide temperatures for the next year by half a degree.

Global Locator

By unhappy coincidence, a typhoon was bearing down on the island at the time of the eruption. The rain-soaked ash caused many roofs to collapse. It formed loose, unstable mud that continued to flow and slide downhill for months, burying towns, wiping out bridges, and ultimately causing more damage than the eruption itself. The two inset photos show the town of Bamban, 30 km away from Mt. Pinatubo, one month (top) and three months after the eruption (bottom).

Despite the damage, the Mt. Pinatubo eruption killed relatively few people. Thanks to early warnings from Earth scientists, most of the area around the volcano had been evacuated. Although 847 people died from the effects of the eruption (including the mudslides afterward), scientists estimate that 5,000 to 20,000 lives were saved by the timely evacuation.

This chapter will explore the processes that lead to volcanic eruptions, the kinds of rocks they form, the ways in which scientists can tell when an eruption might be coming, and the reasons some volcanoes erupt violently and others do not.

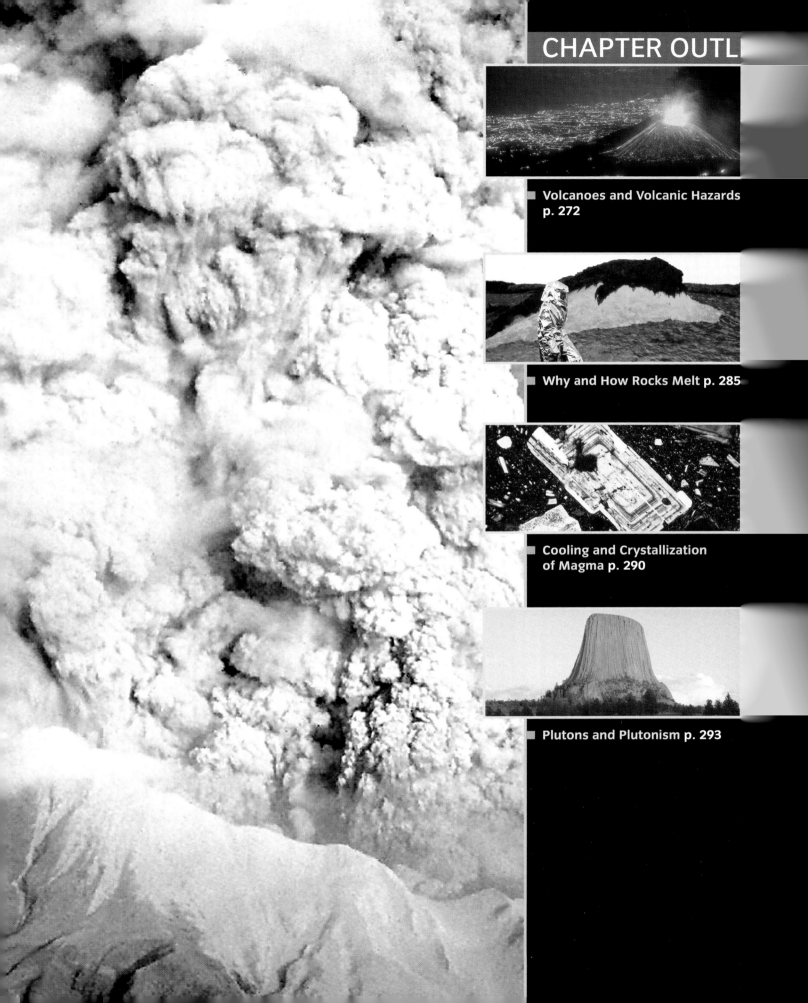

Volcanoes and Volcanic Hazards

LEARNING OBJECTIVES

Identify several different categories of volcanoes.

Explain why stratovolcanoes tend to erupt explosively whereas shield volcanoes tend to erupt nonexplosively.

Describe how volcanic features such as calderas, geysers, and fumaroles arise.

Identify the hazards of volcanoes and the ways in which they can have beneficial effects.

Describe how scientists monitor volcanic activity.

F or many people, the thought of a **volcano** conjures up visions of fountains of **lava** spurting up into the air and pouring out over the landscape (FIGURE 9.1). Although it's true that most volcanoes produce at least some liquid lava, many other types of materials can emerge from volcanoes as well, such as fragments of rock, glassy volcanic ash, and gases. Stored-up gases can cause a volcano to explode, covering the surrounding area with a catastrophic shower of volcanic ash and broken rock (FIGURES 9.1A and 9.1B). Or gases can seep out silently and poison a whole town overnight, as you will discover in the *Case Study*. The different kinds of eruptions and the volcanoes they build have much to do with the physical properties of **magma** that lies at their source. We will begin our discussion by taking a look at some of the different kinds of volcanoes.

> **volcano** A vent through which magma, rock debris, volcanic ash, and gases erupt from Earth's crust to its surface.

> **lava** Molten rock that reaches Earth's surface.

> **magma** Molten rock, with any suspended mineral grains and dissolved gases, that forms when melting occurs in the crust or mantle.

Different eruption styles, different hazards FIGURE 9.1

◀ **A Nonexplosive Eruptions** Hawaii's volcanic eruptions, such as this eruption of Kilauea that began in 1983, are very photogenic and pose minimal danger to humans, although they can cause extensive property damage if lava flows into populated areas.

◀ **B Explosive Eruptions** The May 18, 1980, eruption of Mt. St. Helens in Washington was much more violent than the relatively harmless eruptions of Kilauea in Hawaii. The entire top of Mt. St. Helens was destroyed in a cataclysmic explosion, as seen in these *before* (B) and *after* (C) pictures.

Crater caused by erupt

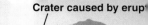

C Here you can see the large crater where the top of the mountain used to be. At least 57 people were killed in the eruption. ▶

Lakes of Death in Cameroon

Two small volcanic lakes in a remote part of Cameroon, a country in central Africa (A), made international news in the mid-1980s when they emitted lethal and invisible clouds of carbon dioxide from deep beneath their surface. The first gas discharge, which occurred at Lake Monoun (not shown) in 1984, asphyxiated 37 people. The second, which occurred at Lake Nyos in 1986, released a highly concentrated cloud of carbon dioxide that killed more than 1700 people. Both occurred at night during the rainy season, both involved volcanic crater lakes, and both are likely to happen again if technological intervention is not successful.

After the incidents, scientists discovered that water in the lakes was stratified, and the bottom layers had huge reservoirs of carbon dioxide dissolved in them. Some minor event—a landslide, or perhaps nothing more than winds at the surface—disturbed the layers, and the stratified column of water turned over. Like a newly opened soda bottle, reduction of pressure on the gas-rich waters allowed the carbon dioxide to bubble to

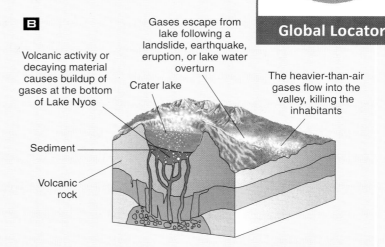

B

Volcanic activity or decaying material causes buildup of gases at the bottom of Lake Nyos

Gases escape from lake following a landslide, earthquake, eruption, or lake water overturn

Crater lake

The heavier-than-air gases flow into the valley, killing the inhabitants

Sediment

Volcanic rock

the surface. In the case of Lake Nyos, approximately 100 million cubic meters of carbon dioxide were released in just two hours (B). Because carbon dioxide is heavier than air, it flowed down the mountainside in a ground-hugging layer that displaced the oxygen that both cattle (C) and people needed to breathe.

The supply of gas at the lake bottom is constantly being replenished from the magma chamber. Scientists estimated that Lake Monoun could experience another violent degassing event within 10 years, and Lake Nyos is at risk within 20 years. It may be possible to siphon off the excess gas by installing a subsurface network of pipes. A prototype system was tested successfully at Lake Monoun in 1992. However, Lake Nyos poses a more complicated problem because of its greater depth and its fragile stratification. Engineers must be very careful not to initiate the very situation that they are trying to prevent.

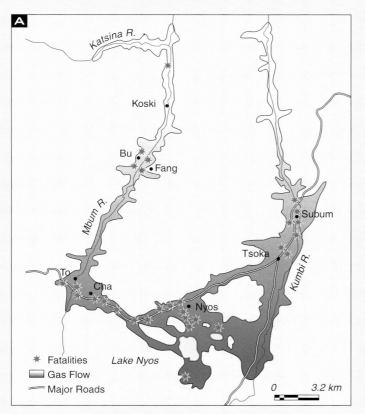

A

Katsina R.

Koski

Bu
Fang

Mbum R.

Subum

Tsoka

Kumbi R.

To
Cha

Nyos

Fatalities
Gas Flow
Major Roads

Lake Nyos

0   3.2 km

C

NATIONAL GEOGRAPHIC

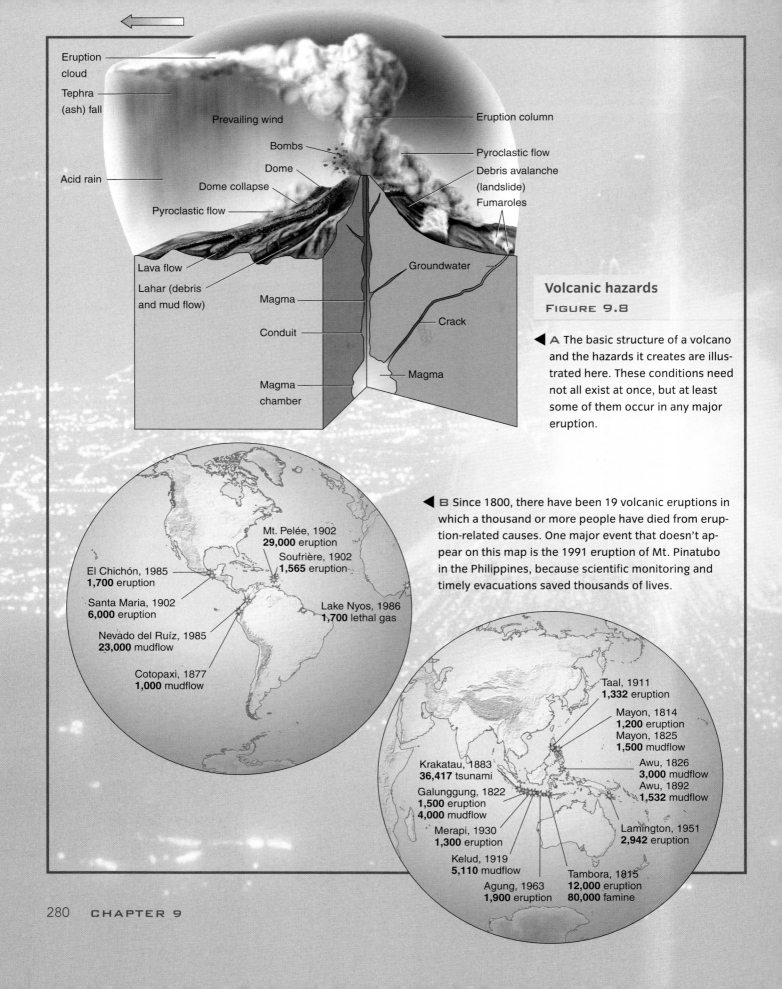

Volcanic hazards
FIGURE 9.8

Eruption cloud

Tephra (ash) fall

Prevailing wind

Acid rain

Bombs

Dome

Dome collapse

Pyroclastic flow

Lava flow

Lahar (debris and mud flow)

Magma

Conduit

Magma chamber

Eruption column

Pyroclastic flow

Debris avalanche (landslide)

Fumaroles

Groundwater

Crack

Magma

◀ **A** The basic structure of a volcano and the hazards it creates are illustrated here. These conditions need not all exist at once, but at least some of them occur in any major eruption.

◀ **B** Since 1800, there have been 19 volcanic eruptions in which a thousand or more people have died from eruption-related causes. One major event that doesn't appear on this map is the 1991 eruption of Mt. Pinatubo in the Philippines, because scientific monitoring and timely evacuations saved thousands of lives.

Mt. Pelée, 1902
29,000 eruption

Soufrière, 1902
1,565 eruption

El Chichón, 1985
1,700 eruption

Santa Maria, 1902
6,000 eruption

Nevado del Ruíz, 1985
23,000 mudflow

Cotopaxi, 1877
1,000 mudflow

Lake Nyos, 1986
1,700 lethal gas

Taal, 1911
1,332 eruption

Mayon, 1814
1,200 eruption
Mayon, 1825
1,500 mudflow

Krakatau, 1883
36,417 tsunami

Galunggung, 1822
1,500 eruption
4,000 mudflow

Merapi, 1930
1,300 eruption

Kelud, 1919
5,110 mudflow

Agung, 1963
1,900 eruption

Awu, 1826
3,000 mudflow
Awu, 1892
1,532 mudflow

Lamington, 1951
2,942 eruption

Tambora, 1815
12,000 eruption
80,000 famine

Volcanoes and climate FIGURE 9.9

◄ **A** The fissure eruption of Laki, a volcano in Iceland, lasted from 1783 to 1784 and was the largest flow of lava in recorded history. Primary and secondary effects killed a third of Iceland's population, and the effects of the eruption were felt widely around the Northern Hemisphere.

▼ **B** In the winter after Laki's eruption, the average temperature in the Northern Hemisphere was about 1°C below normal. In the eastern United States, the decrease was closer to 2.5°C. At the same time, ice cores from Greenland record a dramatic spike in acidity, due to acid precipitation.

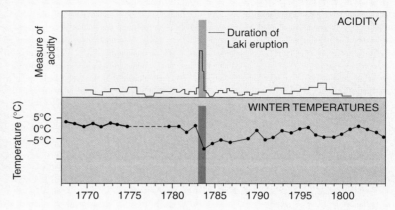

Mudslides have often been a major cause of volcano-related deaths. When volcanic ash mixes with snow at the volcano's summit (or vice versa, when rain falls on recently deposited volcanic ash), it can start a deadly mudflow called a *lahar*. The account at the beginning of the chapter describes a lahar from the eruption of Mt. Pinatubo. Lahars can occur months after the eruption. A related phenomenon is a volcanic *debris avalanche,* in which many different types of material, such as mud, pyroclastic material, and downed trees, are mixed together. A devastating debris avalanche caused much of the damage from the 1980 eruption of Mt. St. Helens.

FIGURE 9.8 summarizes the various primary and secondary hazards from an eruption and shows where the most deadly eruptions of the past two centuries occurred. Note that in many cases the secondary effects, such as the tsunami associated with Krakatau, are responsible for the greatest losses of life.

Tertiary effects

Volcanic activity can change a landscape. Eruptions can block river channels and divert the flow of water. They can dramatically alter a mountain's appearance, as in the 1980 eruption of Mt. St. Helens (Figure 9.1). They can form new land, such as the black sand beaches of Hawaii, which are made of dark pyroclastic fragments, or the volcanic island of Surtsey, which emerged from the ocean near Iceland in 1963 and is composed of both lava flows and pyroclastic cones.

Volcanoes can also affect the climate on a regional and global scale (see **FIGURE 9.9**). Major eruptions can cause toxic and acidic rain, spectacular sunsets, or extended periods of darkness. Sulfur dioxide, a common gaseous emission of volcanoes, forms small droplets or *aerosols*. If they get into the stratosphere, these aerosols spread around the world, absorb sunlight, and cool Earth's surface. An example from recent times is the 1815 eruption of Tambora in Indonesia, which caused three days of near darkness as far away as Australia. The following year was so cool in Europe and North America that it was called "the year without a summer." Farther back in time the eruption of flood basalts, such as the Deccan Traps in India and the Siberian Traps in Russia, may have caused or contributed to several of the mass extinctions that divide geologic periods.

Beneficial effects

Not all the effects of volcanoes are negative, and it is no accident that people choose to live near active volcanoes. Periodic volcanic eruptions renew the mineral content of soils and replenish their fertility—some of the most fertile agricultural lands in the world are adjacent to active volcanoes (see **FIGURE 9.10** on pages 282–283). Volcanism also provides

Global Locator

Asia

Indonesia

Living with danger FIGURE 9.10

Farming villages speckle the slopes of Mt. Merapi, an active volcano on the island of Java, Indonesia. Merapi has a long history of dangerous, often fatal, eruptions, but the fertility of its soils lures farmers to its hazardous slopes. Although volcanic soils cover just 1% of Earth's land surface, they support 10% of the world's population.

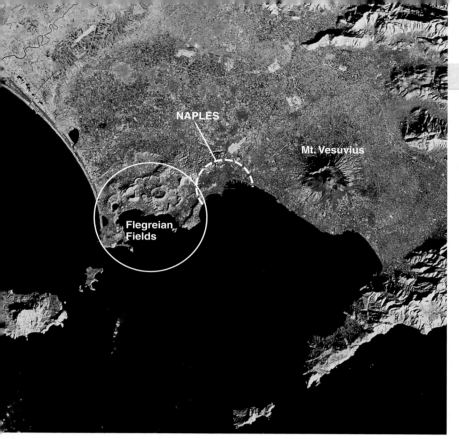

This false-color satellite image shows the area around Mt. Vesuvius (center right) and the Bay of Naples, Italy, a densely populated region. Recent lava flows show up bright red in this image, which records infrared radiation (i.e., heat). Older lavas and volcanic ash show up as shades of yellow and orange. The dark blue and purple region at the head of the bay is the city of Naples. West of Naples lies a cluster of smaller volcanoes called the Flegreian Fields. By comparing successive satellite images, an Earth scientist can detect changes in ground temperature.

geothermal energy and some types of mineral deposits. One rare kind of volcanism brings up diamond-bearing magma from deep in the mantle. All natural gem-quality diamonds on Earth reach the surface by volcanism.

Predicting eruptions

It isn't possible to stop volcanic eruptions, but it is sometimes possible to predict them. The first step in prediction is to identify a volcano as active, dormant, or extinct. An *active* volcano has erupted within recent history; a *dormant* one has not erupted in recent history. A volcano is called *extinct* when it shows no signs of activity and is deeply eroded. Mt. Pinatubo in the Philippines had been dormant for about 500 years prior to its awakening in 1991.

Another important step in prediction is identifying the past eruptive style of a volcano. For example, Mt. Pinatubo is surrounded by thick deposits of pyroclastic material, a sign that the volcano erupted violently in the past. Subduction zone volcanoes such as Mt. St. Helens are more likely to erupt explosively than shield volcanoes and fissure eruptions. The type of rock that has solidified from past eruptions, either silica-rich or silica-poor, also indicates a volcano's style of eruption.

When a volcano shows warning signs of increasing activity, scientists monitor it more closely. Some of these warning signs include changes in the shape or elevation of the ground, such as bulging, swelling, or the formation of a dome. The presence of these features suggests that the underground reservoir of magma is growing. The release of gases can also be a warning sign, as can changes in the temperature of crater lakes, well water, or hot springs. A sudden increase in local seismic activity is also a warning sign. Earth scientists monitor volcanic activity by using tiltmeters to detect bulging, satellite images, and devices that identify gas emissions or changes in the temperature of the ground (FIGURE 9.11).

CONCEPT CHECK STOP

Which kinds of eruptions pose the greatest risk to humans?

Which kinds of eruptions pose comparatively low risk?

Why do pyroclastic flows travel faster and kill more people than do lava flows?

What is the difference between a crater and a caldera?

Why do explosive eruptions occasionally affect global climate?

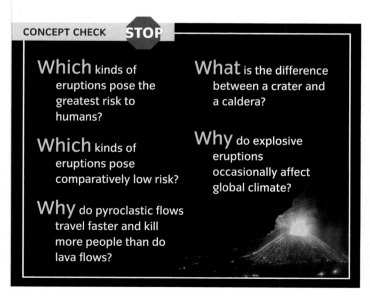

Why and How Rocks Melt

LEARNING OBJECTIVES

Describe how temperature, pressure, and water content affect a rock's melting point.

Define fractional melting.

Identify three properties that distinguish one lava or magma from another.

Underneath every active volcano lies a reservoir of magma, called a *magma chamber*. Understanding volcanism includes understanding how rocks melt to become magma. Fortunately, rocks can be melted artificially as well as naturally (**FIGURE 9.12**). We can thus learn about the behavior of molten rock from laboratory experiments.

At Earth's surface, common types of rock, such as granite and shale, begin to liquefy when heated to a temperature between about 800°C and 1000°C. However, rock (unlike ice, for example) typically consists of many different minerals, each with its own characteristic melting temperature. Thus, we cannot talk about a single melting point for a rock; rather there is a temperature range across which melting occurs. Melting may start at 800°C, but complete melting is commonly attained by about 1200°C. Two other factors beside temperature also strongly affect melting: pressure and the presence of water in the rock.

HEAT AND PRESSURE INSIDE EARTH

If you descend into a mine, it becomes apparent that the farther down you go, the hotter it gets. The rate at which temperature increases with depth, called the **geothermal gradient**, is quite different underneath continental surfaces than it is under the seafloor. Continental crust is thick, and the temperature underneath it increases more gently, starting at about 20°C per km and increasing more rapidly at depth, for an average rate of about 7.0°C per kilometer, reaching 1000°C at a depth of 150 km, which is the base of the lithosphere. Underneath the ocean floor, the rate of increase

> **geothermal gradient** The rate at which temperature increases with depth below Earth's surface.

Molten rock: artificial vs. natural FIGURE 9.12

A In a steel mill, workers heat metal ores to the melting point in order to separate the metal from the surrounding rock.

B An Earth scientist in a protective suit measures the temperature of lava erupting from Mauna Loa, Hawaii. Bright orange, yellow, and white lava is hotter, whereas dull red, brown, and black colors indicate cooler lava.

Geothermal gradient FIGURE 9.13

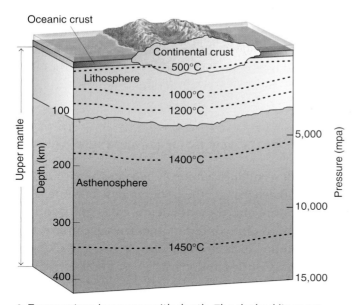

A Temperature increases with depth. The dashed lines are *isotherms,* lines of equal temperature. Notice how the lines "sag" underneath the continental crust, because the rate of increase of temperature is slower there.

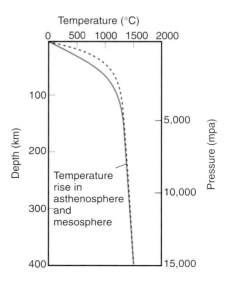

- - - - Temperature rise with depth in oceanic lithosphere
———— Temperature rise with depth in continental lithosphere

B This graph represents the same information as shown in **A**. Earth's surface is at the top, so depth (and pressure) increases as you move down. The dashed curve shows the geothermal gradient under oceans, and the solid curve shows the gradient under continental crust. Note that the two curves merge (and the isotherms become level) below 200 km.

is about twice as rapid. As with the continental crust, the temperature increases more rapidly with depth, for an average rate of 13°C/km, reaching 1000°C at a comparatively shallow depth of 80 km (**FIGURE 9.13**). Below the asthenosphere–lithosphere boundary, heat is transferred by convection and the geothermal gradient becomes more gradual (0.5°C/km), and the temperature difference between suboceanic and subcontinental rock disappears.

As you can see in Figure 9.13, the temperature in the upper mantle is higher than the temperature at which most rocks melt at Earth's surface. Yet the upper mantle is mostly solid. How is this possible?

The answer is that the pressure also rises very dramatically with increasing depth, and increasing pressure causes rock to resist melting (**FIGURE 9.14A**). For example, albite, a common rock-forming mineral (a feldspar), melts at 1104°C at the surface. At a depth

of 100 km, the pressure is 35,000 times greater than it is at sea level. At that pressure, the melting temperature of albite rises to 1440°C, which still slightly exceeds the normal temperature at that depth. Thus, albite remains solid when it is beneath the surface.

The presence of water (or water vapor) in a rock dramatically reduces its melting temperature (**FIGURE 9.14B**). By analogy, as anyone who lives in a cold climate knows, salt can melt the ice on an icy road because a mixture of salt and ice has a lower melting temperature than pure ice. Similarly, a mineral-and-water mixture has a lower melting temperature than the dry mineral alone because water plays the same role as salt in a salt-ice mixture.

The effect of water on the melting of a rock becomes particularly important in subduction zones, where water is carried down into the mantle by oceanic crust, as described in Chapter 7.

Effects of temperature and pressure on melting FIGURE 9.14

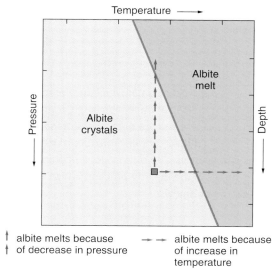

Dry (no H₂O)

A The melting temperature of a dry mineral (albite, in this case) increases at high pressures. A mineral at depth (shown by the small square) can melt in two different ways: either by an increase in temperature (red arrows) or by a decrease in pressure (blue arrows). The latter effect is called *decompression melting,* and it is an important reason why much magma stays molten all the way to the surface.

Temperature →

↑↑ albite melts because of decrease in pressure

→→ albite melts because of increase in temperature

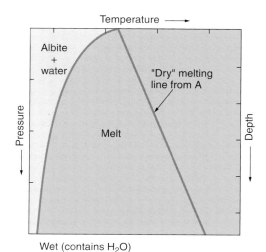

Wet (contains H₂O)

B The melting temperature of a mineral in the presence of water typically decreases as pressure increases. This is exactly the opposite of what happens to dry minerals. Magmas containing dissolved water typically solidify before they reach the surface.

FRACTIONAL MELTING

Because rocks are composed of many different minerals, they melt over a range of temperatures. This means that the boundary between solid and melt is not crisp, but blurry, as in FIGURE 9.15 on page 288. When the temperature rises enough for part of the materials in a rock to melt and part to remain solid, it becomes a **fractional melt**. Only if the temperature continues to increase, or the pressure to decrease, will the rock melt completely. **Fractionation**, an important process that can lead to the development of a diversity of rock types, is caused by fractional melting.

> **fractional melt** A mixture of molten and solid rock.

> **fractionation** Separation of melted materials from the remaining solid matter during the course of melting.

MAGMA AND LAVA

As mentioned earlier, molten rock is called *magma.* When magma reaches the surface, it is called *lava.* A lot of magma never reaches the surface, but instead remains underground, trapped in a magma chamber, until it crystallizes and hardens to igneous rock. We cannot study magma underground in its natural setting, but we can study lava and we can experiment with synthetic magma. From our direct observations of lava, we know that magmas differ in *composition, temperature,* and *viscosity.*

Composition   Most magma is dominated by silicon, aluminum, iron, calcium, magnesium, sodium, potassium, hydrogen, and oxygen—the most abundant elements in the mantle and crust. Oxygen combines with the others to form oxides, such as SiO_2, Al_2O_3, CaO, and H_2O. Silica, or SiO_2, usually accounts for 45% to 75% of the magma, by weight. In addition, a small amount of dissolved gas (between 0.2% and 3% of the magma, by weight) is usually present, primarily water vapor (H_2O) and carbon dioxide (CO_2). Despite their low abundance, these gases strongly influence the properties of magma. The proportion of silica (SiO_2) also has a strong effect on the magma's appearance and

Effect of pressure and temperature on rocks FIGURE 9.15

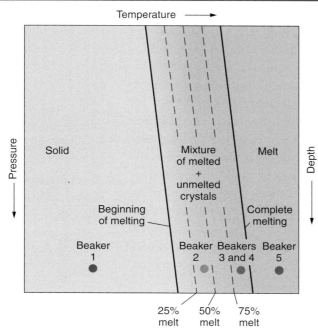

◀ Because almost all rocks contain a mixture of materials, they do not melt all at once, at a single temperature; instead, there is a range of temperatures and pressures in which they contain a mixture of melted and unmelted crystals. The dots on this diagram show the stages in the melting process, which are illustrated further below.

▼ The process of fractional melting can lead to another effect called *fractionation* (illustrated for convenience in a laboratory beaker rather than buried in Earth's mantle).

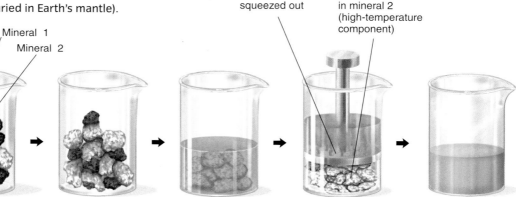

1 The first beaker shows a mixture of two minerals. At a low temperature, both are solid.

2 As the temperature increases, mineral 1 (the dark mineral) begins to melt and as it does so, it dissolves some of mineral 2.

3 Mineral 1 has totally melted and has dissolved a lot of mineral 2 in the process; remainder of mineral 2 remains solid.

4 At a constant temperature, we have mechanically compressed the sample, separating the solid from the melt.*

5 If temperature were to continue to increase, the material in the beaker would eventually become completely melted.

*In the lithosphere, this mechanical separation can occur as a result of the tectonic motion of plates. If the melt now cooled again, we would have two separate deposits, one of mineral 1 plus a little of mineral 2, crystallizing from the melt, and one of pure mineral 2, the solid remnant in Beaker 4.

Here's an interesting question:
• How could the mixture in Beaker 2 become the same as the mixture in Beakers 3 and 4 without any change in temperature?

Plutons and Plutonism

Although they are formed underground, plutonic rocks give rise to some very dramatic geologic formations, known as **plutons** (FIGURE 9.20). Named after Pluto, the Roman god of the underworld, plutons are always *intrusive* bodies, different from the rock that surrounds them. They always originate as magma underground, but they are exposed as plutons at the surface by erosion.

BATHOLITHS AND STOCKS

Plutons are named according to their shapes and sizes. The largest type of pluton is a **batholith** (from the Greek words meaning "deep rock"). Some batholiths exceed 1000 km in length and 250 km in width (FIGURE 9.21 on page 294).

Where they are visible at the surface due to erosion, the walls of batholiths tend to be nearly vertical. This early observation led scientists to believe that batholiths extend downward to the base of Earth's crust. However, geophysical measurements suggest that this perception is incorrect. Most batholiths seem to be only 20 to 30 km thick.

A smaller version of a batholith, only 10 km or so in its maximum dimension, is called a *stock*. In some cases, as shown in Figure 9.20, a stock may be associated with a batholith that lies underneath it.

Most stocks and batholiths are granitic or between granite and diorite in composition. As mentioned earlier, Bowen's hypothesis for the formation of granite batholiths by fractional crystallization of basaltic magma does not stand up to testing. The magma that

pluton Any body of intrusive igneous rock, regardless of size or shape.

batholith A large, irregularly shaped pluton that cuts across the layering of the rock into which it intrudes.

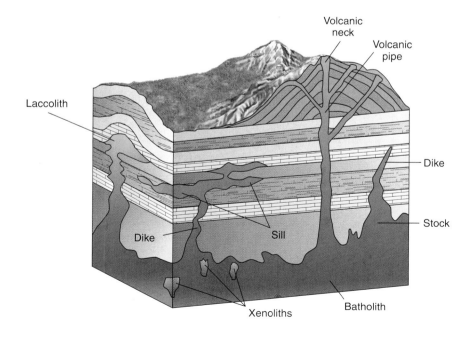

Plutons FIGURE 9.20

This diagram shows the origin of various forms taken by plutons. Note the vertical volcanic necks; sills parallel to the layering in the surrounding rocks; and dikes, which cut across the surrounding rock layers. In every case, the magma intrudes into previously existing rock.

forms batholiths results from extensive fractional melting of the lower continental crust. The heat that causes the melting comes from andesitic magma formed by wet-partial melting of mantle rocks as a result of subduction. Batholiths, such as the Sierra Nevada, were once capped by arcs of andesitic stratovolcanoes. Despite their huge size, the magma bodies that form batholiths migrate upward, squeezing into preexisting fractures and pushing overlying rocks out of their way.

DIKES AND SILLS

Smaller plutons tend to take advantage of fractures in the rock. Two of the most obvious indicators of past igneous activity are **dikes** and **sills** (FIGURE 9.22). A dike forms when magma squeezes into a cross-cutting fracture and then solidifies. If the magma intrudes between two layers and is parallel to them, it forms a sill. Sometimes this intrusion will cause the overlying rock to bulge upward, forming a mushroom-shaped pluton called a *laccolith*. As shown in Figure 9.20, all of these intrusive forms may occur as part of a network of plutonic bodies.

Dikes and sills can be very large. For instance, there is a large and well-known sill-like mass, made of gabbro, in the Palisades, the cliffs that line the Hudson River opposite New York City. The Palisades Intrusive Sheet is about 300 m thick. It formed from multiple charges of magma intruded between layers of sedimentary rock about 200 million years ago. The sheet is visible today because tectonic forces raised that portion of the crust upward, and then the covering sedimentary rocks were largely removed by erosion.

As Figure 9.20 shows, plutonic rocks can also be connected with volcanoes. Beneath every volcano lies a complex network of channels and chambers through which magma reaches the surface. When a volcano becomes extinct, the magma in the channels solidifies into various kinds of plutons. A *volcanic pipe* is the remnant of a channel that originally fed magma to the *volcanic vent*; when exposed by erosion, it is called a *volcanic neck* (Figure 9.22).

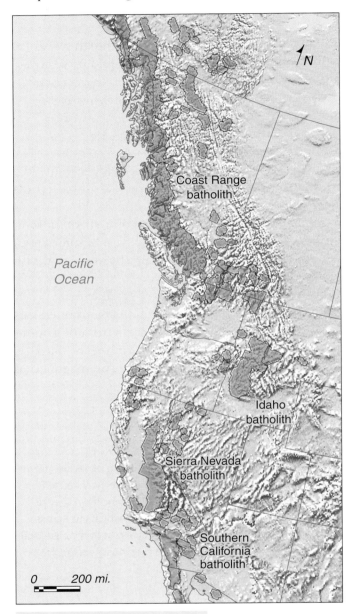

Batholiths FIGURE 9.21

Because batholiths are so immense, we cannot show one in a single photo. However, we can illustrate them on a map. The Coast Range batholith of southern Alaska, British Columbia, and Washington dwarfs the largest batholiths of Idaho and California. Many of the individual stocks and batholiths shown on this map are just the exposed tops of much larger intrusive bodies that lie underground.

CONCEPT CHECK **STOP**

What are some common kinds of plutons?

How do plutons become visible at Earth's surface?

Relative Age

LEARNING OBJECTIVES

Define relative age and numerical age.

Define stratigraphy and identify the four main principles of stratigraphy.

Explain why there are many gaps in the rock record.

Describe how fossils make it possible for Earth scientists to correlate strata in different places.

Through most of human history, when people thought about the origins of features like mountain ranges and oceans, they tended to think in terms of catastrophic events—mountains being thrust up in a single paroxysm or floods that covered the world. This line of thinking came to be called *catastrophism*. Toward the end of the 18th century, a Scottish scientist, James Hutton, made careful studies of the landscape. He noted that erosion of mountains proceeded very slowly and that features he could see in today's sediments were similar to those he could see in ancient rocks that had once been sediments. This led Hutton to the realization of an important principle of science, which later came to be called **uniformitarianism**. One way to express the principle is "the present is the key to the past."

The principle of uniformitarianism provided the first steps toward understanding Earth's history, and further, because things happen so slowly, to the realization that Earth must be incredibly ancient. The first scientific attempts to determine the numerical extent of Earth's history were made a little over two centuries ago. These early Earth scientists speculated that they might be able to estimate the time needed to erode away a mountain range by measuring the rate at which sediment was transported by streams. Their attempts were all underestimates, but the inescapable conclusion supported Hutton—Earth must be millions of years old because of the great thickness of sedimentary rocks. Hutton was so impressed by the evidence that in 1788 he wrote that for Earth there is "no vestige of a beginning, no prospect of an end."

Hutton made careful observations and used the scientific method. His conclusions were at odds with the belief that catastrophes had shaped the landscape and that life on Earth was recent. Such ideas were more in line with those of James Ussher, Archbishop of Armagh in Ireland. In 1658, Ussher published his book, *Annals of the World*, in which he reported his calculations of the date of Creation. He based his work on the histories of the civilizations of the Middle East and the Mediterranean, and on the Bible, and concluded that the day of Creation was Sunday, October 23, 4004 B.C. Sir John Lightfoot, a contemporary of Ussher and Vice-Chancellor of Cambridge University in England, went a step further; he made similar calculations and concluded that the time of day was 9 A.M.

The scientists who followed Hutton agreed with his conclusion that Earth must be very ancient, but they lacked a precise way to determine exactly how long ago a particular event occurred. The only thing they could do was determine the sequence of past events. They could thus establish the **relative ages** of rock layers or other natural features, which means that they could determine whether a particular layer or feature was older or younger than another layer or feature. Relative ages are derived from three basic principles of **stratigraphy**, generally attributed to Hutton and Steno (see Chapter 2), and to the principle of cross-cutting relationships.

uniformitarianism The concept that the processes governing the Earth system have operated in a similar manner throughout Earth's history and that past events can be explained by phenomena and forces observable today.

relative age The age of a rock layer, fossil, or other natural feature relative to another feature.

stratigraphy The science of rock layers and the processes by which strata are formed.

STRATIGRAPHY

In places where you can find large exposed rock formations, such as the Badlands of South Dakota and the American Southwest, you will often see that the rocks have a banded appearance (see **FIGURE 10.1**).

Note, however, that Earth's geographic North and South Poles of rotation do not change position. (Earth's magnetic field is explained in additional detail in Chapter 7.) Scientists are still working out details of how or why **magnetic reversals** happen. The two important points are that the reversal happens quickly by Earth's time standards, and that any iron-bearing mineral in a sedimentary or igneous rock retains or "remembers" the magnetic polarity of Earth at the time that the rock was formed—that is, a change in the magnetic field does not affect already-formed minerals. Through a combination of radiometric dating and magnetic polarity measurements, it has been possible to establish a time scale of magnetic polarity reversals dating back to the Jurassic Period (see Figure 10.16). Still earlier reversals are a subject of ongoing research.

Correlation on the basis of magnetic reversals differs from other stratigraphic correlation methods. One magnetic reversal looks just like any other in the rock record. When evidence of a magnetic reversal is found in a sequence of rocks, the problem lies in knowing which of the many reversals it actually represents. When a continuous record of reversals can be found, starting with the present, it is simply a matter of counting backward. But if not, the technique must be combined with stratigraphic and radioactive dating techniques (see the *Case Study*). Chapter 7 discusses how magnetic polarity studies played a crucial role in the development of plate tectonic theory.

> **magnetic reversal** The period of time in which Earth's magnetic polarity reverses itself.

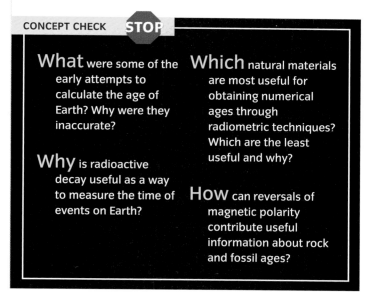

CONCEPT CHECK STOP

What were some of the early attempts to calculate the age of Earth? Why were they inaccurate?

Why is radioactive decay useful as a way to measure the time of events on Earth?

Which natural materials are most useful for obtaining numerical ages through radiometric techniques? Which are the least useful and why?

How can reversals of magnetic polarity contribute useful information about rock and fossil ages?

The Age of Earth

LEARNING OBJECTIVES

Explain why the oldest rocks are not necessarily the same age as the planet.

Explain why scientists currently believe Earth is about 4.56 billion years old.

Throughout this book, we mention examples of actual rates of Earth processes. This would not be possible without the numerical dates obtained through radiometric dating and other numerical age methods. In fact, more than any other contribution by scientists, the ability to determine numerical dates has changed the way humans think about the world and the immensity of Earth's history.

Now that we know how to determine the numerical ages of rocks, can we determine Earth's age? It's not as easy as you might think. The continual recycling of Earth's surface by erosion and plate tectonics means that very few, if any, remnants of

Earth's oldest rocks FIGURE 10.17

The Acasta gneiss in northern Canada was formed 4.0 billion years ago. It is the most ancient body of rock so far discovered on Earth.

Earth's original crust remain. Of the many radiometric dates obtained from Precambrian rocks, the oldest is about 4.0 billion years (see FIGURE 10.17). Although no rocks older than this have been found, an individual mineral grain from a sedimentary rock in Australia has been dated to 4.4 billion years, so it is conceivable that igneous rocks older than 4.0 billion years may someday be located.

However, because the oldest mineral grain came from a sedimentary rock, there must have been a period of rock formation and erosion before sedimentation, so scientists understand that there is still a gap between the age of the mineral grain and the age of Earth. There is strong evidence that Earth formed at the same time as the Moon, the other planets, and meteorites. Through radiometric dating, it has been possible to determine the ages of meteorites and of "moon dust" brought back by astronauts. The *Apollo* astronauts found rocks and individual grains of Moon dust that are pieces of the Moon's original crust. Such rocks are abundant because the Moon system has been tectonically much less active than the Earth system.

Meteorite ages are especially valuable because some meteorites have remained virtually unaltered since the formation of the solar system. Melting and other types of alteration will reset the radiometric clock. However, some meteorites, such as the Allende Meteorite, which fell to Earth in the Mexican state of Chihuahua on February 8, 1969, belong to a rare category called *carbonaceous chondrites*, which, as far as we can tell, contain unaltered material from the formation of the solar system. It is carbonaceous because it contains tiny amounts of carbon (about 3 parts per 1000). Some of the carbon is in chemical compounds called *amino*

A cosmic interloper FIGURE 10.18

The Allende Meteorite, which fell to Earth in Mexico, is one of the most famous meteorites in history. Note the white spots on the meteorite. Some of these inclusions, which are slightly older than the black carbonaceous material around them, are more than 4.6 billion years old, making them the oldest objects of any kind ever found on Earth.

acids—organic components that are essential for life. The dark, fine-grained part of the meteorite is mostly olivine, with a few flecks of metallic iron and some carbon. The clumps of white material are oxides of calcium and aluminum and are thought to be among the first matter to condense from the gas cloud from which the solar system formed (FIGURE 10.18). The white clumps are older than Earth itself.

The ages of many of these objects from the solar system cluster closely around 4.55 billion years. Planetary scientists therefore conclude that Earth, and indeed the Sun's entire planetary system, formed at that time.

Today, more than two centuries after Hutton, it is widely agreed that Earth's age is approximately 4.55 billion years. When will it cease to exist? Astronomers tell us that billions of years in the future, the Sun will become a red giant, at which point it will expand and engulf Earth. However, Hutton is still correct in one sense: Earth's history is profound and evidence here on Earth shows no prospect of an end.

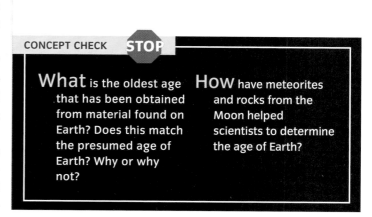

CONCEPT CHECK **STOP**

What is the oldest age that has been obtained from material found on Earth? Does this match the presumed age of Earth? Why or why not?

How have meteorites and rocks from the Moon helped scientists to determine the age of Earth?

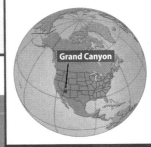

Global Locator

Tapeats Sandstone

Angular unconformity

Grand Canyon Supergroup

Standing at the rim of the Grand Canyon, you can see more than 2 billion years of Earth's history preserved in the rocks. All three kinds of unconformities can be found here. The upper layers in this photograph were all deposited during the Paleozoic Era. Some of the contacts between different-colored parallel strata are disconformities, and they record the rising and ebbing of seas over this part of the North American continent. Below the Tapeats Sandstone (arrow) you can see an angular unconformity, which also represents a major time gap between Precambrian rocks (deposited about 825 million years ago) and Cambrian rocks (deposited less than 542 million years ago). Finally, there is a nonconformity (not visible in the photo) between the lowermost sedimentary layer of the Grand Canyon Supergroup, the Bass Limestone, and the Vishnu Formation, a foundation of metamorphic and igneous rocks that once lay at the base of an ancient mountain range that eroded away long before the Rocky Mountains came into existence.

As an Earth science student, you owe it to yourself not to stop at the rim, as most tourists do, but to descend to the river level and get a closeup look at 2 billion years of Earth's history. You might be lucky and see trilobite tracks, like these, in the Tapeats Sandstone. The tracks were made when trilobites (like the one shown in Figure 10.10) extended their legs sideways, pulled in mud, then, under the safety of their hard shells, picked over the mud for food.

NATIONAL GEOGRAPHIC

SUMMARY

1 Relative Age

1. Earth scientists study the chronologic sequence of natural events, that is, their **relative age**. Relative age is derived from **stratigraphy**, the study of rock layers and how those layers are formed.

2. There are four basic principles used to determine relative ages. **Strata**, or sedimentary rock layers, are horizontal when they are deposited as water-laid sediment (*law of original horizontality*). Strata accumulate in sequence, from the oldest on the bottom to the youngest on the top (*principle of stratigraphic superposition*). A rock stratum is always older than any feature, such as a fracture, that cuts across it (*principle of cross-cutting relationships*). Finally, rock strata extend outward horizontally; they may thin or pinch out at their farthest edges, but they generally do not terminate abruptly unless cut by a younger rock unit (*principle of lateral continuity*).

3. **Numerical age**, the exact number of years of a natural feature or event, is more difficult to determine. One difficulty that arises is that the sequence of strata in any particular location is not necessarily continuous in time. An **unconformity** is a break or gap in the normal stratigraphic sequence. It usually marks a period during which sedimentation ceased and erosion removed some of the previously laid strata. The three common types of unconformities are nonconformities, angular unconformities, and disconformities.

4. **Correlation** of strata is the establishment of the time equivalence of strata in different places. **Fossil** assemblages, usually consisting of hard shells, bones, and plant material, have been the primary key to correlation of strata across long distances. The study of fossils and the record of ancient life on Earth is called **paleontology**. The principle of *faunal and floral successions* (animals and plants, respectively) is the stratigraphic ordering of fossil assemblages.

2 The Geologic Column

1. The **geologic column**, a *stratigraphic time scale,* is a composite diagram that shows the succession of all known strata, arranged in chronological order of formation, based on fossils and other age criteria.

2. The geologic column is divided into several different units of time, called **eons**, **eras**, **periods**, and **epochs**. The majority of Earth's history is divided into two eons, in which fossils are very rare or nonexistent. Those eons, each spanning several hundred million years, are the *Archean* and *Proterozoic*. The third and most recent eon, the *Phanerozoic*, is the only eon in which fossils are abundant. Very dramatic changes in fossil assemblages occur between the three eras of the Phanerozoic Eon—the *Paleozoic, Mesozoic,* and *Cenozoic*—which were separated by major extinction events. The earliest period of the Paleozoic Era is especially important, as this was a time of unique diversification. This period is known as the *Cambrian Explosion*. Except for the very end of the Proterozoic Eon, rocks formed before the Cambrian Period cannot be differentiated by the fossil record.

3 Numerical Age

1. Determining the age of Earth and of events that have happened on Earth has long been of interest to scientists. Different hypotheses were proposed by Halley, Joly, Darwin, Lord Kelvin, and several other prominent scientists of the later 1800s and early 1900s. Though many of these hypotheses were found to be incorrect, each was an important step in eventually finding a method of numerical dating.

2. **Radioactivity** is the process in which an element spontaneously transforms itself into another *isotope* of the same element, or into a different element, through the release of particles and heat energy. The *radioactive decay* of isotopes of chemical elements provides a basis for radiometric dating, which gives values for the numerical ages (age in years) of rock units and thus values for numerical dates of past events. Because radioactive decay is not influenced by chemical processes or by heat and high pressure in Earth, it is an extremely accurate gauge of numerical age.

3. **Radiometric dating** is based on the principle that in any sample containing a radioactive isotope, half of the atoms of that isotope will change to daughter atoms within a specific length of time, called the **half-life**. (The *proportion* of parent atom decay during a unit of time is always the same.) Radioactive isotopes with a long half-life, such as uranium, are most useful for dating rocks. Carbon-14, which has a much shorter half-life, is most useful for dating organic materials of relatively recent origin (less than 70,000 years).

4. Though radiometric dating is primarily useful for igneous rocks, a complementary technique called *magnetic polarity dating* works for sedimentary rocks, too.

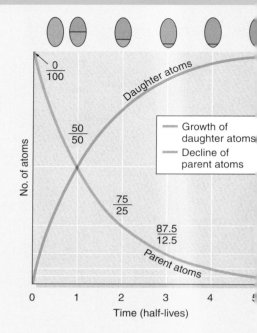

Magnetic polarity dating involves paleomagnetism, the study of reversals in Earth's magnetic field. As yet, **magnetic reversals**, or periods of time in which Earth's magnetic polarity reverses itself, are not fully understood.

4 The Age of Earth

1. Through measures of numerical age, it has become clear that most of Earth's history took place in Precambrian time. The oldest Earth rocks discovered are about 4.0 billion years old.

2. Earth is not a good place to look for the oldest rocks in the solar system. Earth's surface has been subjected to a lot of erosional activity. This has reset some radiometric clocks and destroyed the earliest fragments of the crust. Samples from the Moon and from meteorites indicate that the solar system formed about 4.56 billion years ago, and by inference this is also the age of Earth.

KEY TERMS

- uniformitarianism p. 304
- relative age p. 304
- stratigraphy p. 304
- numerical age p. 309

- unconformity p. 309
- paleontology p. 311
- correlation p. 312
- geologic column p. 312

- radioactivity p. 318
- half-life p. 320
- radiometric dating p. 320
- magnetic reversal p. 325

CRITICAL AND CREATIVE THINKING QUESTIONS

1. Do the principles of stratigraphy apply on the Moon in the same way as they do on Earth? Bear in mind that the processes operating on the Moon have been very different from those on Earth. If you had to determine the relative age of features on the Moon, based entirely on satellite photographs, what would you look for and how might you proceed?

2. Check the area in which you live to see whether there is an excavation—perhaps one associated with a new building or road repair. Visit the excavation and note the various layers, the paving (if the excavation is in a road), and the soil below the surface. Is any bedrock exposed beneath the soil?

3. How old are the rock formations in the area where you live and attend college or university? How can you find out the answer to this question?

4. Choose one of the geologic periods or epochs listed in Figure 10.9A and find out all you can about it: How are rocks from that period identified? What are its most characteristic fossils? Where are the best samples of rocks from your chosen period found?

5. Do some research to determine the ages of the oldest known fossils. What kind of life-forms were they?

What is happening in this picture ?

This skier is hauling a sled past a cliff face on Ellesmere Island, Canada.

Why do you think the rock strata in the background tilt at such a steep angle?

Why are they wavy instead of straight?

1. A _____ is the age of one rock unit or geologic feature compared to another.
   a. numerical age
   b. relative age
   c. radioactivity age
   d. stratigraphic age

2. The principle of cross-cutting relationships says that _____.
   a. waterborne sediments are deposited in nearly horizontal layers
   b. a sediment or sedimentary rock layer is younger than the layers below it and older than the layers that lie above
   c. a rock unit is older than a feature that disrupts it, such as a fault or igneous intrusion
   d. a sediment or sedimentary rock layer is older than the layers below it and younger than the layers that lie above

3. The _____ states that waterborne sediments are deposited in nearly horizontal layers.
   a. law of superposition
   b. principle of faunal succession
   c. law of original horizontality
   d. principle of cross-cutting relationships

4. In a conformable sequence, _____.
   a. each layer must have been deposited on the one below it without any interruptions
   b. there must not be any depositonal gaps in the stratigraphic record
   c. Both a and b are correct.
   d. None of the above is true.

5. An unconformity represents _____.
   a. a gap in the stratigraphic record
   b. a period of erosion or no deposition
   c. Both a and b are correct.
   d. None of the above is true.

6. On this illustration, label each unconformity as one of the following:
   a. nonconformity
   b. angular unconformity
   c. disconformity

7. Fossils found in strata _____.
   a. are the records of ancient life
   b. allow the correlation of strata separated by many miles
   c. have been useful to geologists in creating the geologic column
   d. All of the above statements are correct.

8. The three eras that make up the Phanerozoic Eon are the _____.
   a. Hadean, Archean, and Proterozoic
   b. Paleozoic, Mesozoic, and Cenozoic
   c. Triassic, Jurassic, and Cretaceous
   d. Pliocene, Pleistocene, and Holocene

9. The most distinctive changes in the fossil record occur across the boundaries between _____.
   a. periods
   b. eras
   c. epochs
   d. disconformity

10. The dinosaurs were dominant during _____.
    a. the Cenozoic Era
    b. the Mesozoic Era
    c. the Paleozoic Era
    d. the Precambrian time

Bald cypress trees

Anchiceratops

Edmonto

Cockroach

Magnolia

Ground beetle

Plant beetle

11. Label the two decay sequences depicted in this illustration as either alpha emission or beta decay. For each decay sequence, also label the following:

    a. parent nucleus
    b. alpha-particle
    c. daughter nucleus
    d. beta-particle

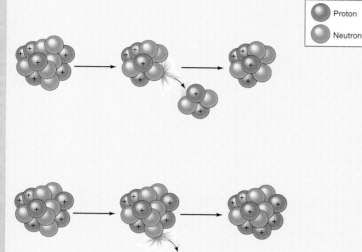

Proton

Neutron

12. Potassium-40 is a naturally occurring radioisotope that decays to Argon-40 and is common in many rocks of the continental crust. The half-life of Potassium-40 is 1.3 billion years. Assuming no contamination, what would be the age of a sample that contained a 1:1 ratio of Potassium-40 to Argon-40?

    a. 1.3 billion years
    b. 650 million years
    c. 2.6 billion years
    d. 325 million years

13. If the sample indicated above showed evidence that it had been heated by contact with a more recent lava flow, what would be the likely error in the determined age?

    a. The sample would appear too young.
    b. The sample would appear too old.
    c. Rocks are a closed system; there would be no error.

14. A gravel deposit containing an important hominid tooth fossil is found in a field location in northern Ethiopia. The gravel deposit has fragments of volcanic rocks dated at 3.75 million years ±0.1 by Potassium–Argon dating (K–Ar) and is known from stratigraphy to be younger than a 2.8-million-year ±0.04 (K–Ar) volcanic ash deposit. Sediments interlayered with the fossil-bearing gravels have good magnetic signals and have a normal polarity. Given the figure in the second column for the region, what is the most likely date for the hominid fossil?

    a. between 4.2 and 3.3 million years
    b. between 3.6 and 3.3 million years
    c. exactly 3.75 million years
    d. younger than 2.8 million years

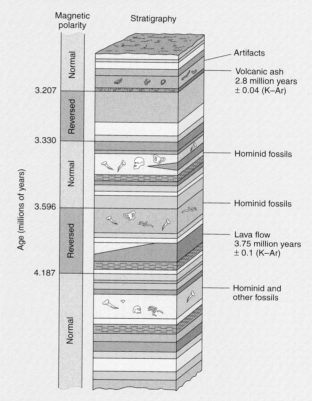

15. Earth is not considered a good place to look for the oldest rocks in the solar system because _____.

    a. contamination from atmospheric tests of nuclear weapons have contaminated the crust of Earth
    b. Earth's magnetic field interferes with the radiometric clocks in most igneous rocks
    c. melting has reset radiometric clocks in the rocks of Earth's crust
    d. the earliest crustal rocks have been destroyed by geologic activity
    e. All of the above are true.
    f. Answers c and d are correct.

A Brief History of Life on Earth

The history and diversity of life, from this ferocious *Tyrannosaurus rex* (on exhibit at the American Museum of Natural History) to the humblest microbe, is very much a part of the planet's history. Throughout this book, you will find numerous examples of interactions between the biosphere and other parts of the Earth system. Plants and microorganisms accelerate the mechanical weathering of rocks and formation of soil. The skeletons of plankton sink to the bottom of the sea, where they form sediments that eventually turn into limestone. Land plants form coal, and animals leave fossils that paleontologists use to reconstruct the history of life. Biologic processes, including the transpiration of plants and respiration of animals, regulate the very air that we breathe.

This interaction also works in the opposite direction: The Earth system affects the course of life on this planet. Earth's atmosphere and hydrosphere provided a habitat in which life could develop and prosper. The forms that life takes are governed by the need to survive in a particular environment—the scalding waters around a mid-ocean rift, the arid land of a desert, the humid climate of a tropical rainforest, all of which result from natural processes.

The study of Earth science is thus inseparable from the study of life on Earth. In this chapter, we trace the story of life from its beginnings. We examine in greater detail how life has been affected by its interactions with the atmosphere, hydrosphere, and lithosphere and how, in turn, life has shaped the Earth system.

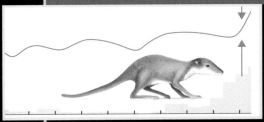

The Ever-Changing Earth

Throughout this book we have seen evidence that Earth is a place of constant change and that negative feedbacks seem to moderate the changes so that Earth continues to be habitable. The most obvious changes, and the ones most readily demonstrated, are the rearrangements of the continents and oceans as a result of plate tectonics. Refer back to Figure 1.11 and review the dramatic continental rearrangements that have occurred over the past 500 years. There is abundant evidence to demonstrate that plate motions have been operating for at least 2 billion years and probably even longer, but the exact locations of continents and oceans during the Precambrian is still uncertain and the focus of much research. In this chapter, we will present evidence for change in Earth's life zone (a summary of the changes is shown in **FIGURE 11.1**).

The changing Earth FIGURE 11.1

From left to right, this diagram illustrates some of the major events in the history of Earth's surface environment during its first 4.6 billion years.

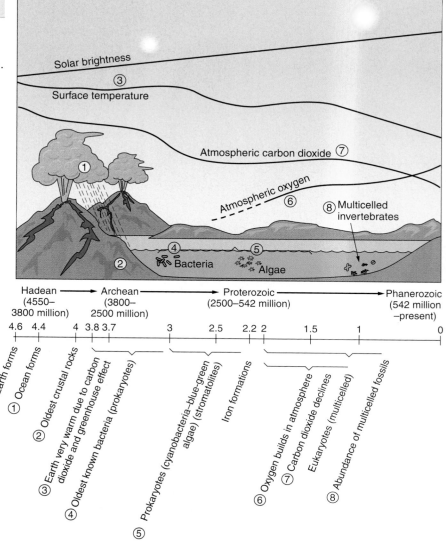

A This pea plant, like all green, leafy plants, produces oxygen by the process of photosynthesis.

Photosynthesis FIGURE 11.2

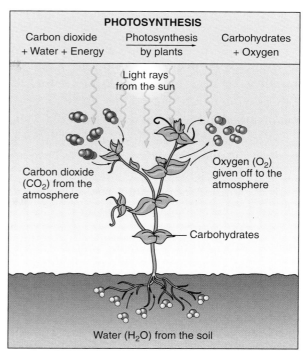

B The photosynthetic reaction combines carbon dioxide (CO_2) and water (H_2O) to make carbohydrates (molecules containing C, H, and O), which the plant needs to grow. The reaction produces oxygen molecules (O_2). The plant releases the oxygen into the atmosphere through pores in its leaves. The reaction does not happen spontaneously, but requires energy from sunlight; that is why it is called *photosynthesis* (*photo-,* meaning "light").

CHANGES IN THE ATMOSPHERE AND HYDROSPHERE

Like other parts of the Earth system, the atmosphere has changed through time (see Figure 11.1). Compared to the Sun, which is representative of the raw materials from which Earth and the other planets of our solar system formed (Chapter 17), Earth contains less of some volatile elements, such as nitrogen, argon, hydrogen, and helium. These elements were lost when the envelope of gases or *primary atmosphere* that surrounded early Earth was stripped away by the solar wind or by meteorite impacts, or both. Little by little, Earth generated a new, *secondary atmosphere,* by volcanic outgassing of volatile materials from its interior.

Volcanic outgassing continues to be the main process by which volatile materials are released from Earth—although it is now going on at a much slower rate. The main chemical constituent of volcanic gases (as much as 97% by volume) is water vapor, with varying amounts of nitrogen, carbon dioxide, and other gases. In fact, the total volume of volcanic gases released over the past 4 billion years or so accounts for the present composition of the atmosphere, with one extremely important exception: oxygen. As you can see in Figure 11.1, Earth had virtually no oxygen in its atmosphere more than 4 billion years ago, but the atmosphere is now approximately 21% oxygen.

Traces of oxygen were probably generated in the early atmosphere by the breakdown of water molecules into oxygen and hydrogen by ultraviolet light (a process called *photodissociation*). Although this is an important process, it doesn't even come close to accounting for the present high levels of oxygen in the atmosphere. Almost all of the free oxygen now in the atmosphere originated through **photosynthesis** (FIGURE 11.2).

photosynthesis A chemical reaction whereby plants use light energy to induce carbon dioxide to react with water, producing carbohydrates and oxygen.

Oxygen is a very reactive chemical, so at first most of the free oxygen produced by photosynthesis was combined with iron in ocean water to form iron oxide–bearing minerals. The control exerted by oxygen on iron dissolved in the ocean was discussed in Chapter 3 (see *What an Earth Scientist Sees*, page 76). Evidence of the gradual transition from oxygen-poor to oxygen-rich ocean water is preserved in seafloor sediments. The minerals in seafloor sedimentary rocks that are more than about 2.5 billion years old contain *reduced* (oxygen-poor) iron compounds. In rocks that are less than 1.8 billion years old, *oxidized* (oxygen-rich) compounds predominate. The sediments that were precipitated during the transition contain alternating bands of red (oxidized iron) and black (reduced iron) minerals. These rocks were deposited as chemical sediments and are called *banded iron formations* (Chapter 3). Because ocean water is in constant contact with the atmosphere, and the two systems function together in a state of dynamic equilibrium, the transition from an oxygen-poor to an oxygen-rich atmosphere also must have occurred during this period.

Along with the buildup of molecular oxygen (O_2) came an eventual increase in ozone (O_3) levels in the atmosphere. Because ozone filters out harmful ultraviolet radiation, this made it possible for life to flourish in shallow water and finally on land. This critical stage in the evolution of the atmosphere was reached between 1100 and 542 million years ago. Interestingly, the fossil record shows an explosive diversification of life-forms 542 million years ago (the beginning of the Phanerozoic Eon).

Oxygen has continued to play a key role in the evolution and form of life. Over the last 200 million years, the concentration of oxygen has risen from 10% to as much as 25% of the atmosphere, before settling (probably not permanently) at its current value of 21% (see **FIGURE 11.3**). This increase has benefited humans and all other mammals, because mammals are voracious oxygen consumers. Not only do we require oxygen to fuel our high-energy, warm-blooded metabolism, our unique reproductive system demands even more. An expectant mother's used (venous) blood must still have enough oxygen in it to diffuse through the placenta into her unborn child's bloodstream. It would be very difficult for any mammal species to survive in an atmosphere of only 10% oxygen.

Scientists cannot yet be certain why the atmospheric oxygen levels increased, but they have a hypothesis, and it illustrates the interactions between all parts of the Earth system. First, note that photosynthesis is only one part of the oxygen cycle. The cycle is completed by decomposition, in which organic carbon combines with oxygen and forms carbon dioxide. But if organic matter is buried as sediment before it fully decomposes, its car-

The oxygen content of the atmosphere FIGURE 11.3

Over the last 200 million years, oxygen levels in the atmosphere have increased markedly. The rise of mammals, though aided by the demise of the dinosaurs 65 million years ago, may also have resulted partly from the plentiful oxygen supply.

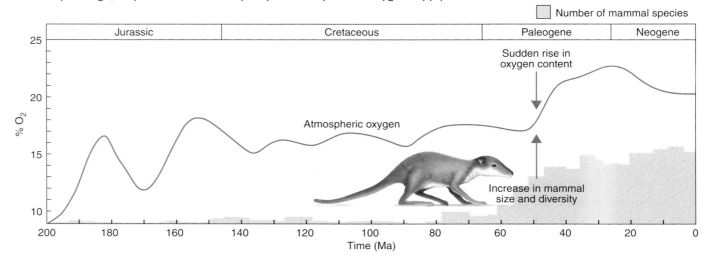

| PERMIAN | ⊘ 251 | TRIASSIC | ⊘ 200 |

Pangaea forms . . . Conifers appear, colonize uplands   First dinosaurs appear . . . First true mammals
First beetles and flies . . . Mammal-like reptiles (therapsids)   95%   80%

A When a living coelacanth was caught in the Indian Ocean in 1938, it created a sensation because not only its species but its entire order, the Crossopterygians, had been believed to be extinct. The first fish to haul themselves onto land may have been relatives of the coelacanth. Today's coelacanths, like this one photographed near the Comoro Islands, are exclusively deep-sea creatures.

B Another candidate for the first fish to make the transition to land is the lungfish. Unlike the coelacanth, today's lungfish still survive for short periods on land without water.

Pioneer fish? FIGURE 11.17

Fishes and amphibians

The phylum of greatest interest to most of us, because humans belong to it, is the *chordates*. These are animals that have at least a primitive version of a spinal cord (called a *notochord*). Like all the other animal phyla, chordates can be found in fossils of the Cambrian Period, although they are relatively inconspicuous. The earliest so far discovered is *Haikouichthys,* a jawless fish akin to a hagfish or a lamprey that lived 525 million years ago—only 17 million years after the beginning of the Cambrian explosion.

Jawed fish arrived next. With jaws, fish could lead a predatory lifestyle and grow to much larger sizes; the jawless fish had been limited to filtering food out of the water or dredging it from the sea floor. Among the first large jawed fish were sharks and ray-finned fish. The earliest known intact shark fossil is 409 million years old.

The first fish to venture onto land may have been a member of an obscure order called *Crossopterygii,* or lobe-finned fish (see FIGURE 11.17). These fish had several features that could have enabled them to make the transition to land. Their lobe-like fins contained all the elements of a quadruped limb. They had internal nostrils, characteristic of air-breathing animals. As fish, they had already developed a vascular system that was adequate for life on land.

The first terrestrial chordates, amphibians, have never become wholly independent of aquatic environments, because they have not developed an effective method for conserving water. To this day they retain permeable skins. They have also never really met the reproductive requirement for life on land. In most amphibian species, the female lays her eggs in water and the male fertilizes them there after a courtship ritual. The young (e.g., tadpoles) are fish-like when first hatched. Just like the seedless plants, amphibians have kept one foot (figuratively speaking) on land and one foot in the water. They originated in the Devonian Period.

Life in the Phanerozoic Eon   353

Gymnosperms and ferns dominate . . . Pangaea breaks apart
Huge increase in size and diversity of dinosaurs . . . Mammals remain small . . . First bird (*Archaeopteryx*)

THE MESOZOIC ERA

The Paleozoic Era closed with the extinction of trilobites and a great many other marine creatures. An estimated 96% of all living species disappeared at the end of the Permian Period in the greatest mass extinction in Earth's history. We will return to the question of mass extinctions at the end of this chapter, but the cause of the extinction remains an enigma that has yet to be solved. The era that followed the Paleozoic, the Mesozoic Era, commenced with all the major landmasses on Earth joined together in the supercontinent Pangaea (see Figure 1.11). Then, in the middle of the era, the supercontinent began to split apart in a process that continues today.

Whatever the cause of the great extinction at the end of the Permian, it did not seem to greatly affect plant life on land. Gymnosperms, which had appeared in the Devonian Period, dominated the plant world during the Triassic and Jurassic Periods. Gymnosperms, however, have an important liability. The male cell-

carrier, the pollen, is spread through the air. This is extremely inefficient. What chance does a pollen grain in the air have of finding a female cell? Eventually, at the beginning of the Cretaceous period, **angiosperms**—flowering plants—found a more efficient solution. For a small incentive, such as nectar or a share of the pollen, insects deliver pollen directly from one flower to another, or from one part of a plant to another. After pollination, the plant develops a seed in much the same way as gymnosperms. In many cases, birds and other animals help distribute the seeds by eating the plant's seed-bearing fruits and distributing the seeds in their feces. Flowering plants still dominate land plants today.

> ■ **angiosperm** A flowering, or seed-enclosed, plant.

Reptiles, birds, and mammals

The Carboniferous and Permian Periods were times when amphibians were abundant. Most families died out in the end-of-Permian extinction. One branch of the amphibians,

Early birds and mammals FIGURE 11.18

A The skeletons and teeth of *Archaeopteryx* were very similar to those of dinosaurs. However, the very detailed impressions of feathers in this fossil identify *Archaeopteryx* as a bird.

B Discovered in 2002 in China, this shrew-sized *Eomaia scansoria* specimen is the oldest known fossil of a placental mammal (that is, a mammal that gives live birth). It lived 125 million years ago, during the height of the dinosaur age.

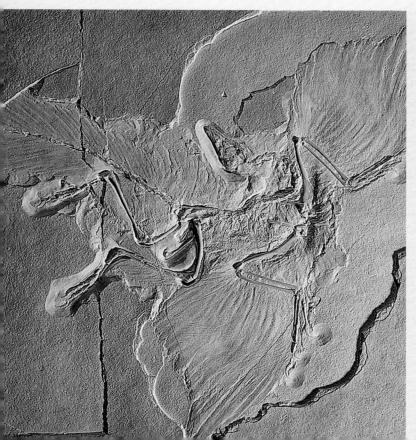

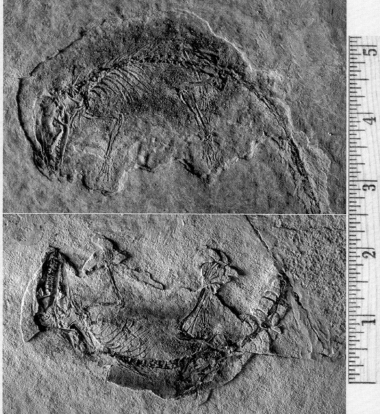

145

CRETACEOUS

⊘
65

Angiosperms (flowering plants) appear and proliferate...
Dinosaurs reach largest sizes ever...Marsupials and placental mammals...Meteorite impact causes mass extinction

70%

however, evolved into reptiles—the first fully terrestrial animals. Reptiles were freed from the water by evolving an egg that contained amniotic fluid for the young to grow in and by developing a watertight skin. These two evolutionary advances enabled them to occupy many terrestrial niches that the amphibians had not been able to exploit because of their need to live near water. The amniotic egg led to an explosion in reptile diversity, much as the evolution of the jaw had done for fishes. Moving out of the Mississippian and Pennsylvanian swamps, some colonized the land, some moved back into the water, and a few took to the air. Not only did reptiles greatly increase in diversity during the Jurassic Period, they also grew to tremendous size. The dinosaurs were the largest land animals that ever lived, possibly ranging up to 100 metric tons in weight and 35 m in length.

Birds first appeared near the end of the Jurassic Period, and they are now considered to be direct descendants of the dinosaurs. An early bird, *Archaeopteryx* (see FIGURE 11.18A), would have been classified as a dinosaur if not for the discovery that it had feathers. Even before *Archaeopteryx*, vertebrates had made the transition to the air in the form of pterosaurs—flying reptiles with long wings and tails. Pterosaurs are not considered by most paleontologists to be true dinosaurs, although they were very closely related. The detail of the transition from reptiles to birds remains one of the most highly controversial topics in paleontology today.

Mammals are descended from a class of "mammal-like reptiles" that existed as long ago as the Permian Period. The transition from reptiles to mammals is well understood, though perhaps not quite as well understood as the transitions from fishes to amphibians. It is difficult to pick out a single mammalian adaptation comparable to the jaws of fish, the eggs of reptiles, or the feathers of birds. During the Cretaceous Period, mammals certainly did not outcompete the dinosaurs; they survived by being small and inconspicuous (FIGURES 11.18B and 11.18C).

THE CENOZOIC ERA

Evidence suggests that an accidental catastrophe—a giant meteorite impact—was at least partially responsible for the environmental changes that wiped out the great reptiles and 70% of all other species, too, at the end of the Mesozoic Era. The departure of dinosaurs gave mammals a chance to grow larger and to diversify. In the Cenozoic Era, mammals have also benefited from the atmosphere's high oxygen level, conducive to a fast metabolism. Unlike reptiles, whose brain sizes have not grown (relative to their body size), mammals have continued to evolve toward larger brain size throughout the Cenozoic Era. This may be one key to mammalian success, because it has enabled them to diversify their lifestyles to a much greater extent than reptiles ever could have.

The last frontier for plants—the dry steppes, savannahs, and prairies—was not colonized until the Paleogene Period, when grasses evolved. This process involved the assistance of animals, in particular the great grazing herds that lived on all continents except Antarctica.

C The shape of *Eomaia*'s claws shows that it was a climber that could grasp tree branches. It is now known that small, tree-dwelling mammals began to experiment with gliding flight as early as 130 million years ago-almost as early as the first birds began flying.

65
TERTIARY
1.8
Present
QUATERNARY
⊘ Possible mass extinction due to human impact on environment?

Grasses evolve . . . Mammals increase in diversity and size . . . Grazing mammals . . .
High oxygen levels enable large brains, fast metabolism . . . First hominids . . . Ice ages

Genus *Homo* . . .
Modern Humans

Lucy's relatives FIGURE 11.19

◀ **A** This drawing by Michael Rothman depicts a mother and child of the species *Australopithecus afarensis*. These human-like individuals lived together in small groups, formed lasting bonds with mates, and looked after their children through infancy.

B This 3.3-million-year-old fossil is the skull of an *A. afarensis* baby, who probably looked much like the child in the drawing. ▶

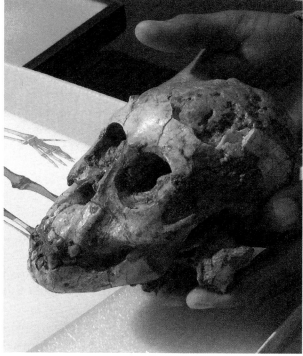

C From footprints like these, preserved in soft volcanic mud, scientists know that australopithecines walked upright on two feet. This 70-m trail includes the footprints of two adults and possibly a child, stepping in the footprints of one of the adults. To the right are footprints of an extinct three-toed horse. ▶

The human family

From a human point of view, it is surely true that the most remarkable thing that has happened during the Cenozoic Era is the emergence of our own species. Charles Darwin was often accused of believing that humans are descended from the apes. In fact, the family of humans, *Hominidae,* and the family of apes, *Pongidae,* are both descended from an earlier common ancestor that was neither human nor ape. The emergence of humans is one of the most complex and controversial fields in paleontology, in part because of the paucity of the fossil record and the lack of transitional forms. But it is clear that **hominids**—human-like organisms—are a very recent evolutionary development. The first hominid that was clearly *bipedal* (walked upright) was *Australopithecus,* of which the famous fossil "Lucy" is an example (see FIGURE 11.19). These hominids were only about 1.2 m in

Amazing Places: The Burgess Shale

In 1909, the American paleontologist Charles Walcott discovered the world's preeminent site for Cambrian fossils, a rock formation called the Burgess Shale, about 90 km away from Banff, British Columbia. It is a treasure trove for specimens such as this trilobite (**A**), as well as more exotic organisms such as the five-eyed Opabinia (**B**) and others. It was only in the 1970s that paleontologists realized that many of the organisms found there are unrelated to any living animals. They are not only extinct species, but members of extinct phyla. Exploring the Burgess Shale is like exploring the evolutionary paths that life could have taken—but didn't.

Many of the animals preserved in the Burgess Shale were soft-bodied. Such creatures fossilize only under very special circumstances. This is how scientists reasoned that it happened: The Burgess fauna lived in shallow, oxygen-rich waters atop an algal reef with steep sides (**B**). Periodically, mudslides would sweep some unlucky animals off the side of the reef and deposit their bodies at the base (**C**). At that time the deep waters were still so oxygen-poor that no microorganisms existed at that depth to decompose them. The fine

silt particles from the mudslide encased the bodies and hardened into shale, preserving a highly detailed imprint of the animals. The *Waptia fidensis* fossil in *What an Earth Scientist Sees* is an example.

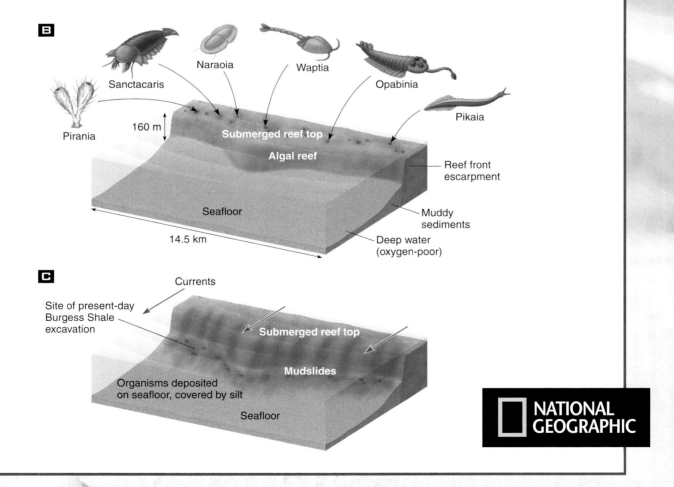

Developed by the Wildlife Conservation Society, the footprint is a measure of population density and land use. The areas least impacted by humans are shown in green.

The human footprint on Earth's surface FIGURE 11.23

**The human footprint
on planet Earth**

Most impacted

Least impacted

SUMMARY

1 The Ever-Changing Earth

1. The history of life is intertwined with that of the atmosphere and hydrosphere. Earth's early atmosphere consisted primarily of water vapor, carbon dioxide, and nitrogen, the products of outgassing from volcanoes.

2. Oxygen gradually accumulated over more than a billion years, primarily through the process of **photosynthesis**. Photosynthesis, a reaction by which plants convert carbon dioxide and water into carbohydrates, is by far the most important source of oxygen in our atmosphere.

3. The buildup in oxygen was accompanied by a buildup of ozone in the upper atmosphere. Earth's land surfaces were probably uninhabitable until the ozone layer began blocking harmful ultraviolet radiation from the Sun.

4. Living plants produce oxygen. When dead organisms decay, the bacteria that decompose them consume oxygen and release carbon dioxide. This cycle maintains the balance of oxygen in the atmosphere and hydrosphere. Nevertheless, atmospheric oxygen levels have fluctuated dramatically over geologic time. One reason this can occur is the burial of organic matter before it has completely decomposed. Plate tectonics can accelerate the rate of burial of organic matter, and thereby affect the composition of the atmosphere.

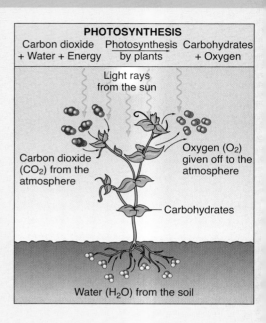

PHOTOSYNTHESIS

Carbon dioxide + Water + Energy → Photosynthesis by plants → Carbohydrates + Oxygen

Light rays from the sun

Carbon dioxide (CO_2) from the atmosphere

Oxygen (O_2) given off to the atmosphere

Carbohydrates

Water (H_2O) from the soil

2 Early Life

1. The minimum requirements for life are a *metabolism* and a means of *replication*. All life on Earth is based on a small set of carbon-based building blocks called *amino acids*. Also, all living organisms use **DNA**, a biopolymer, to copy genetic information. However, different organisms have evolved different kinds of metabolism, and this is one reason that life is able to flourish in a great variety of environments.

2. The fundamental unit of life is the **cell**, which has two fundamental varieties: **prokaryotes** and **eukaryotes**. Eukaryotic cells enclose their DNA within a *membrane,* whereas prokaryotic cells lack a well-defined nucleus.

3. Biologists have classified all living organisms into three **domains**—bacteria, *Archaea,* and eukaryotes. The first two domains are prokaryotes, which are exclusively single-celled organisms. Prokaryotes evolved long before eukaryotes did. Eukaryotes evolved at least 1.4 billion years ago and can be either one-celled or multicelled.

4. The most ancient fossils are about 3.5 billion years old, although chemical evidence of the presence of life is preserved in rocks considerably older than this. The best pieces of evidence for early life are fossilized *stromatolites,* thick mats of one-celled organisms that can still be found in some places today. Most other fossils of early life are microscopic, and consequently they are hard to find and difficult to interpret. The earliest known fossils of large multicellular organisms date back 635 million years.

3 Evolution and the Fossil Record

1. As a result of **evolution**, all present-day organisms are descendants of different kinds of organisms that existed in the past. The mechanism by which evolution occurs is **natural selection**, in which well-adapted individuals tend to have greater survival and more breeding success, and thus pass along their traits to later generations.

2. New **species** may emerge slowly, as the environment changes gradually over time, or quickly, when a beneficial *mutation* appears or when a population moves to a new habitat. It is not clear which process is more common.

3. Organisms can be preserved as **fossils** in many different ways. The body may be kept essentially intact, or its shape may be preserved while minerals replace its contents. Some organisms leave behind evidence, such as *molds* of footprints or **trace fossils**, without leaving behind any body parts.

4 Life in the Phanerozoic Eon

1. The Cambrian Period, at the beginning of the Phanerozoic Eon, was a time of explosive diversification of life-forms. Several factors may have contributed to the *Cambrian explosion*: the emergence of multicellular life (just before the Cambrian Period), the protective ozone layer, the rise in oxygen levels, and the ability to form shells.

2. All of the extant **phyla** in the animal **kingdom**, as well as several extinct phyla, emerged during the Cambrian Period. All of the Cambrian fauna were aquatic, and some had hard internal or external skeletons. In order to colonize land, plants and animals had to develop a means of *structural support,* an *internal aquatic environment,* and a way of *exchanging gases* with the atmosphere. For sexual reproduction, a *moist environment* was also essential.

3. The earliest land plants were seedless plants, such as ferns. The next major event in the plant kingdom was the emergence of **gymnosperms** or naked-seed plants, which flourished during the age of dinosaurs. **Angiosperms**, or flowering plants, developed last, with a much more efficient reproductive system facilitated by insect pollination.

4. The earliest land animals were arthropods (a phylum that includes insects). Fish first ventured onto land during the Devonian Period, and eventually gave rise to the amphibians. These, however, were limited in their geographic range because they still required water to spawn in.

5. Reptiles evolved a watertight skin and eggs that could be incubated outside of water. This freed them from dependence on watery environments. Both mammals and birds descended from reptiles, by different evolutionary pathways.

(continued)

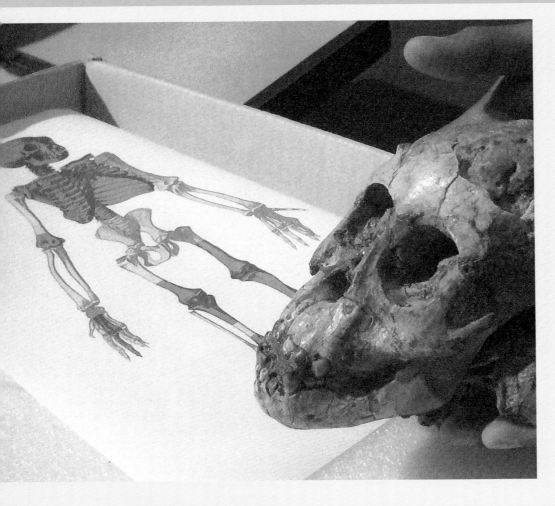

6. Mammals have existed since the Jurassic Period. It was apparently the disappearance of the dinosaurs, along with a rise in atmospheric oxygen, which gave mammals the opportunity to diversify and increase in size. **Hominids** are a very recent family of mammals, which appeared within the last 3.9 million years. Anatomically modern humans appeared only 30,000 years ago. Scientists are not certain whether Neandertals, which died out around the same time, were a distinct species.

7. Several geologic time periods are delineated by **mass extinctions**. The causes of mass extinctions are not thoroughly understood, but most scientists now believe that the end-of-Cretaceous extinction, which killed the dinosaurs, resulted from a meteorite impact. Other mass extinctions may have been caused by volcanism.

KEY TERMS

CRITICAL AND CREATIVE THINKING QUESTIONS

1. Recall from Chapter 1 (see Table 1.1) that Earth and Venus are so similar in size and overall composition that they are almost "twins." Why did these two planets evolve so differently? Why is Earth's atmosphere rich in oxygen and poor in carbon dioxide, whereas the reverse is true on Venus? What would happen to Earth's oceans if Earth were a little bit closer to the Sun?

2. One classical criticism of the theory of evolution by natural selection was the paucity of transitional species. However, *Archaeopteryx* is such a species. In what ways did it resemble its dinosaur predecessors? In what ways was it like or unlike modern birds?

3. Another common criticism of Darwin's theory was its alleged inconsistency with the Biblical account of Creation. Yet Darwin himself was a Christian, and many scientists since Darwin have had no difficulty in reconciling the evidence for evolution with their religious beliefs. Investigate how they have done so. Do you personally feel that evolution by natural selection conflicts with your religious beliefs?

4. What do you think might have happened to mammals if the end-of-Cretaceous extinction had not wiped out the dinosaurs?

What is happening in these pictures ?

In this fossil bird and this fossil frog, found in the Messel Oil Shale near Darmstadt, Germany, the animals' soft tissue, such as feathers and skin, are exceedingly well preserved. (Note the faint imprint of the frog's skin around its bones.)

- Why do you think the soft body parts were preserved in this case, even though they are seldom found in most fossils?

1. Which of the following statements best describes changes in Earth's environment over the last 4 billion years?
   a. Solar brightness and oxygen generally increase while surface temperature and atmospheric CO_2 decrease.
   b. Solar brightness and oxygen generally decrease while surface temperature and atmospheric CO_2 increase.
   c. Solar brightness and surface temperature generally increase while oxygen and atmospheric CO_2 generally decrease.
   d. Solar brightness and surface temperature generally decrease while oxygen and atmospheric CO_2 generally increase.

2. Photosynthesis is fundamentally important to life on Earth. On this illustration, label the input and output of the photosynthetic process using the following terms:

   carbon dioxide CO_2 light rays

   water H_2O carbohydrates

   oxygen O_2

3. Although scientists cannot be certain why the atmospheric oxygen levels increased, they hypothesize that atmospheric oxygen has increased over time _____.
   a. as a continued increase in photosynthetic organisms
   b. because of increased rates of seafloor spreading
   c. because organic matter is buried before it decomposes, reducing the formation of carbon dioxide (CO_2)
   d. All of the above statements are correct.

4. The minimum requirements for life are _____.
   a. photosynthesis and a means of replication
   b. mobility and a means of replication
   c. metabolism and a means of replication
   d. mobility and metabolism

5. There are two cells in these photographs. One is from a prokaryotic and the other from a eukaryotic organism. Label structures in each of the cells with the following terms, and then label each cell either *prokaryotic* or *eukaryotic*.

   organelles   cell membrane   nucleus

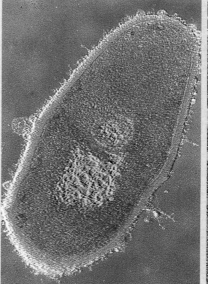 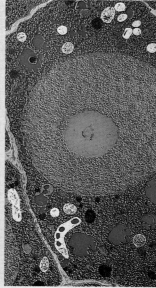

6. _____ is an example of anaerobic metabolism.
   a. Chemosynthesis   d. Both a and b are correct.
   b. Photosynthesis   e. Both a and c are correct.
   c. Fermentation

7. Natural selection is the process by which _____.
   a. individuals migrate to environments better suited to survival of the species
   b. individuals adapt to their environment over time
   c. well-adapted individuals pass on their survival advantages to their offspring
   d. All of the above are correct.

8. As a mechanism for evolution, natural selection requires the passing on of specific traits from one generation to the next. These specific traits may reflect _____ passed on through an organism's _____.
   a. genetic mutations/metabolism
   b. genetic mutations/DNA
   c. genetic mutations/RNA
   d. metabolism/DNA
   e. metabolism/RNA

9. How are organisms preserved as fossils within the rock record?
   a. through the preservation of the organism in substances such as sap, tar, or ice
   b. through minerals replacing the contents of the dead organism (mineralization)
   c. by the leaving of impressions, molds, or footprints
   d. through mummification in especially dry environments
   e. All of the above are methods through which organisms could be preserved as fossils.

10. Which of the following hypotheses do scientists think best explains the explosion of life-forms represented by the Cambrian fossil record?
    a. Sexual reproduction began in the early Cambrian, leading to a greater diversity of living organisms.
    b. An oxygen-rich atmosphere allowed for the metabolism of larger organisms.
    c. Predation began in the early Cambrian, leading to the development of many more organisms with shells and skeletons that might be better preserved in the fossil record.
    d. Extreme climatic changes at the end of the Proterozoic Eon led to accelerated evolutionary responses in surviving organisms.
    e. All of the above hypotheses are considered equally valid by scientists today.

11. Which of the following is not a requirement for a living organism to survive on land?
    a. structural support
    b. a means of locomotion
    c. an internal aquatic environment
    d. a mechanism of exchanging gases with the air
    e. if sexually reproducing, a moist environment for the reproductive system

12. Vascular plants exchange gas with the air by _____.
    a. diffusion through adjustable openings in the leaves called *stomata*
    b. osmosis involving oxygenated water in the root system
    c. photosynthesis
    d. All of the above are correct.

13. Reptiles were freed from the water by _____.
    a. evolving an egg that contained amniotic fluid for the young to grow in
    b. developing a watertight skin
    c. developing a warm-blooded metabolism
    d. Both a and b are correct.
    e. Both b and c are correct.

14. _____ is probably the first species of our own genus.
    a. *Australopithecus*
    b. *Homo erectus*
    c. *Homo neandertalensis*
    d. *Homo sapiens*

15. Mass extinctions on Earth may be correlated with _____.
    a. massive eruptions of flood basalts
    b. the impact of massive meteorites
    c. glaciation
    d. Both a and b are correct.
    e. Both b and c are correct.

The Oceans

Imagine a dry Earth, devoid of water. Viewed from an orbiting spacecraft, a dry Earth would no longer have a bluish color, the land would lack a vegetation cover, and no clouds would obscure the surface. Earth would appear as desolate as the Moon or Mars.

Circling Earth, we would see several vast, interconnected basins, each floored with oceanic crust and rimmed with continental crust. If these huge basins were slowly filled with water, the scene would be transformed. First, the rising water would fill the deepest parts of the basins, creating a number of shallow seas. These seas would then merge to form an ever-larger ocean that would eventually creep up the slopes of the continental margins. With the ocean basins filled to capacity, more than two-thirds of Earth's surface would be covered by water and Earth, viewed from space, would be a blue planet unique in the solar system.

Under the ocean, beyond the continental slopes, lies the remote world of the deep ocean floor. We now have devices for mapping the ocean floor—with its volcanoes and canyons—and teams of marine scientists have explored it. The romanticist in each of us may regret that beliefs and legends built up through more than 3000 years of human history—monsters, mermaids, strange and threatening sea gods, fabled cities and castles that sank into watery depths—have vanished. But in their place, we are beginning to find a new world rich in unimaginable undersea life.

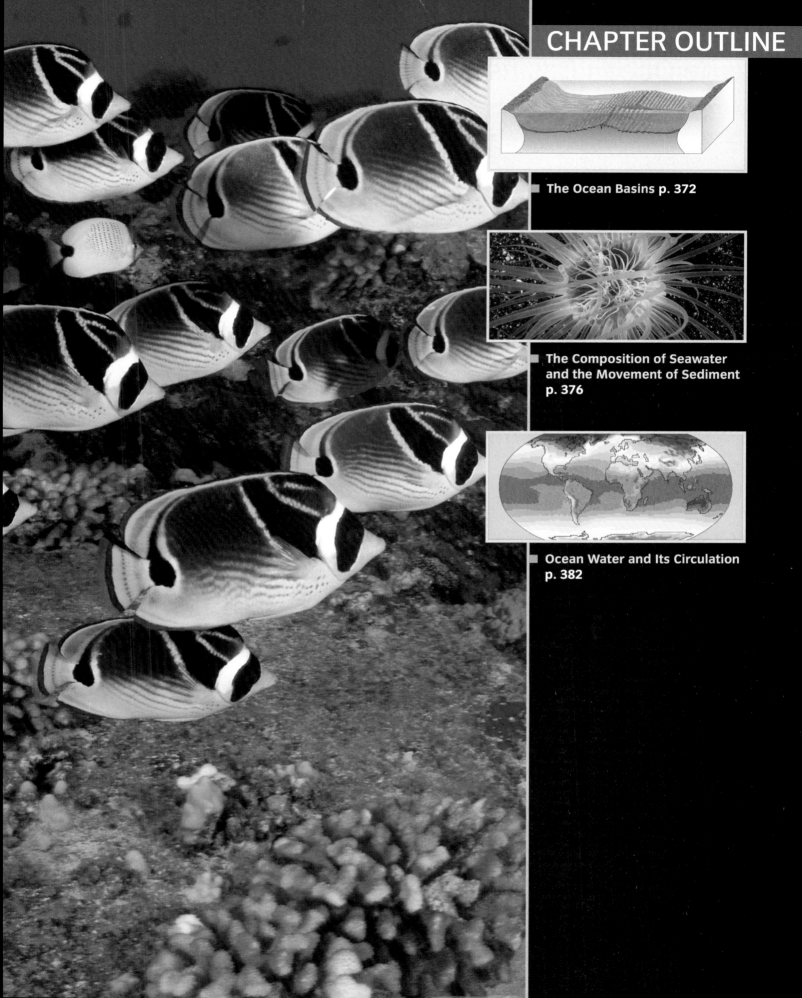

The Ocean Basins

LEARNING OBJECTIVES

Discuss the age and origin of the oceans.

Describe the distribution of water and land over the globe.

Explain how ocean depth is measured.

Identify the important features of ocean basins, including the midoceanic ridge, the axial rift, and the abyssal plains.

ater has been present on Earth's surface since the planet's earliest days. In fact, we do not have any definitive evidence that there was ever a time when Earth lacked water. The oldest rock so far discovered on Earth is gneiss that was once contained in *sedimentary strata*. Since these strata are about 4.0 billion years old and were originally deposited in water, we can be reasonably certain that the ocean formed sometime between 4.56 billion years ago (when Earth was formed) and 4.0 billion years ago. The most ancient mineral grain so far discovered on Earth is a grain of zircon from a sandstone in Western Australia (see Chapter 11). That grain is 4.4 billion years old, and the oxygen isotopes it contains record contact with water, so it is possible that there was an ocean of some sort as much as 4.4 billion years ago. But regardless of when the earliest ocean began to form, just where the water in it came from is still an open question (**FIGURE 12.1**).

The origin of the oceans FIGURE 12.1

Scientists do not yet agree on where Earth's water came from.

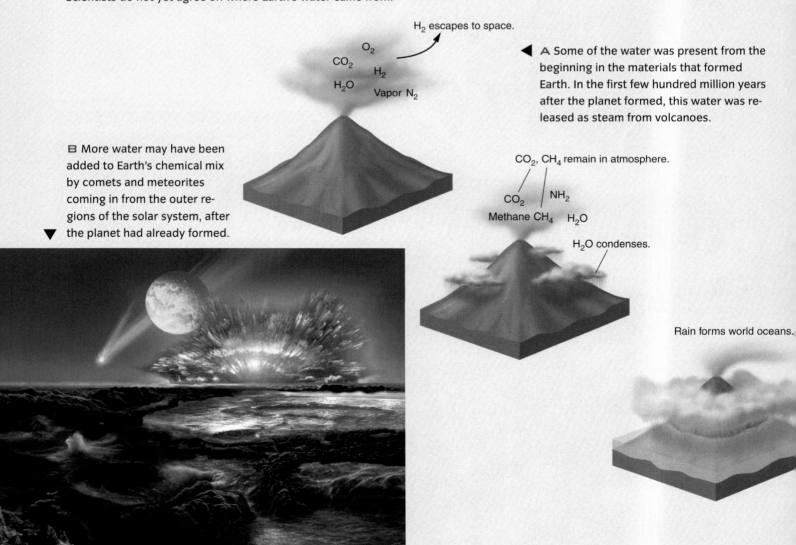

H_2 escapes to space.

O_2
CO_2
H_2
H_2O
Vapor N_2

A Some of the water was present from the beginning in the materials that formed Earth. In the first few hundred million years after the planet formed, this water was released as steam from volcanoes.

B More water may have been added to Earth's chemical mix by comets and meteorites coming in from the outer regions of the solar system, after the planet had already formed.

CO_2, CH_4 remain in atmosphere.

CO_2 NH_2
Methane CH_4 H_2O

H_2O condenses.

Rain forms world oceans.

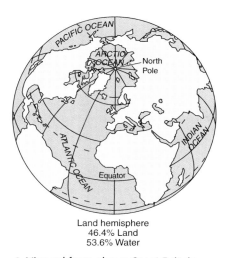 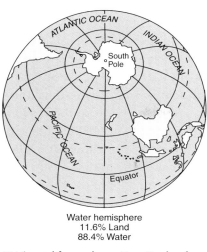

Land hemisphere
46.4% Land
53.6% Water

Water hemisphere
11.6% Land
88.4% Water

A Viewed from above Great Britain, over 46% of the Northern Hemisphere is land.

B Viewed from above New Zealand, over 88% of the Southern Hemisphere is water.

OCEAN GEOGRAPHY

Seawater covers 71% of Earth's surface. The land that covers the remaining 29% is unevenly distributed. This uneven distribution is especially striking when we compare two views of the globe: one from a point directly above Great Britain and the other from a point directly above New Zealand (FIGURE 12.2). The uneven distribution of land and water plays an important role in determining the paths along which water circulates in the open ocean and its marginal seas. We will look at ocean circulation later in the chapter.

Most of the water on our planet is contained in four huge interconnected **ocean basins**. The Pacific, Atlantic, and Indian oceans are connected with the Southern Ocean, the body of water south of 50°S latitude that completely encircles Antarctica. (The fifth major ocean, the Arctic Ocean, is considered an extension of the North Atlantic.) Collectively, these four bodies of water, together with a number of smaller ones, are often referred to as the *world ocean.*

> **ocean basins**
> Regions of Earth's crust that are covered by seawater.

The smaller water bodies connected with the Atlantic Ocean include the Mediterranean, Black, North, Baltic, Norwegian, and Caribbean Seas; the Gulf of Mexico; and Baffin and Hudson Bays. The Persian Gulf, Red Sea, and Arabian Sea are part of the Indian Ocean, while among the numerous marginal seas of the Pacific Ocean are the Gulf of California, the Bering Sea, the Sea of Okhotsk, the Sea of Japan, and the East China, South China, Coral, and Tasman Seas. All these seas and gulfs vary considerably in shape and size; some are almost completely surrounded by land, whereas others are only partly enclosed. Each owes its distinctive geography to plate tectonics. This ongoing process, described in Chapter 7, has created many small basins both in and next to the major ocean basins.

DEPTH AND VOLUME OF THE OCEANS

Before the 20th century, we knew little about the depth of the oceans. Sailors determined the depth of water by lowering a weighted hemp line, or a strong wire, until it hit the bottom. This technique was effective and relatively rapid in shallow water, but it could take 8 to 10 hours to recover a weighted wire in water thousands of meters deep. By the close of the 19th century, about 7000 measurements had been made in water more than 2000 m (6500 ft) deep, and fewer than 600 had been made in water more than 9000 m (29,500 ft) deep.

In the 1920s, ship-borne acoustical instruments called *echo sounders* were developed to measure ocean depths. An echo sounder generates a pulse of sound and accurately measures the time it takes for the echo bouncing off the seafloor to return to the instrument. Because we know the speed of sound traveling through water, we can work out the depth of the water beneath the ship.

Topography of the ocean basin By far the longest chain of mountains on Earth—some 64,000 km in length—is hidden from our view. That's because it's underwater, twisting and branching in a complex pattern through the ocean basins. Plate tectonics has given the ocean floor a distinctive topography, and the features of ocean basins are quite different from those of continents. Much of the oceanic crust is less than 60 million years old, and the oldest oceanic crust so far discovered is only 200 million years old. By contrast, the great bulk of the continental crust is over 1 billion years old.

FIGURE 12.3 shows the important relief features of ocean basins. The basin is divided in about half by a *midoceanic ridge* of submarine hills. In the center of the ridge, at its highest point, is a narrow, trench-like feature called the *central rift valley* (also known as the *axial rift*). The location and form of this rift suggest that the crust is being pulled apart along the line of the rift. On either side of the midocean ridge is a broad, deep plain, known as an *abyssal plain.*

Over the past 80 years, the oceans have been crossed many thousands of times by ships carrying echo sounders. As a result, we have been able to map the topography of the seafloor and the depth of the water lying directly above it in considerable detail for all but the most remote parts of the ocean basins. The greatest ocean depth yet measured (11,035 m; 36,205 ft) lies in the Mariana Trench, near the island of Guam in the western Pacific. This is more than 2 km (6500 ft) farther below sea level than Mt. Everest rises above sea level. Based on recent satellite measurements, the average depth of the sea is about 4500 m (14,760 ft), compared to an average height of the land of only 750 m (2460 ft). The present volume of seawater is about 1.35 billion cubic kilometers (324 million mi^3); more than half this volume resides in the Pacific Ocean. We say *present* volume, because the amount of water in the ocean fluctuates over thousands of years, mainly because of the growth and melting of continental glaciers (see Chapter 6).

Figure 12.3 shows a symmetrical ocean-floor model that fits the North Atlantic, South Atlantic, Indian, and Arctic Ocean Basins well. These oceans have *passive continental margins* that have not been subjected to strong tectonic and volcanic activity during the last 50 million years. This relative inactivity is due to the fact that the continental and oceanic lithospheres that join at a passive continental margin are part of the same lithospheric plate and move together, away from the axial rift.

But unlike the symmetrical ocean-floor model of the North Atlantic, the margins of the Pacific Ocean Basin have deep offshore oceanic trenches (see *What an Earth Scientist Sees*). We call these trenched ocean-basin edges *active continental margins*. Here, oceanic crust is being bent downward and is sinking under continental crust, creating trenches and inducing volcanic activity.

Ocean basins
FIGURE 12.3

This schematic block diagram shows the main features of ocean basins. It applies particularly well to the North and South Atlantic Oceans.

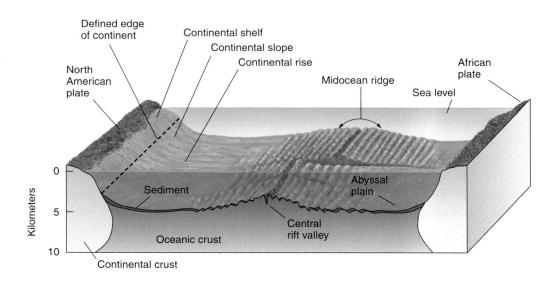

Undersea Topography

These images of the ocean floor were constructed from radar data that measured the surface height of the ocean very precisely. Deeper regions are shown in tones of purple, blue, and green, while shallower regions are shown in tones of yellow and reddish brown. In **A**, you can see a prominent ring of trenches in the western Pacific Basin. An Earth scientist would recognize that these are *subduction* trenches, created, as shown in **B**, when oceanic crust is being bent downward and forced under continental crust. In **C**, you can see the mid-Atlantic ridge, which is the result of seafloor spreading (**D**). Data were acquired by the U.S. Navy Geosat satellite altimeter.

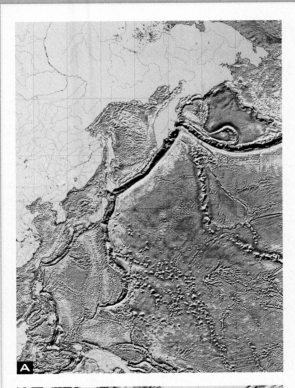

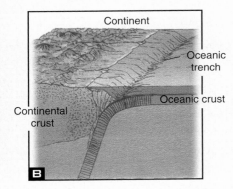

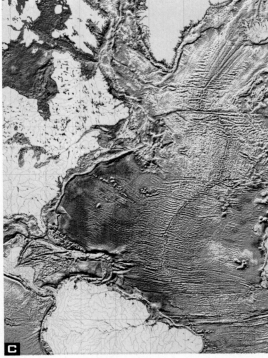

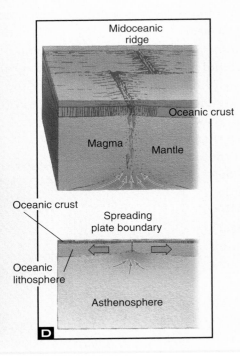

Here's an interesting question:

CRITICAL THINKING

- In **Figure A**, there is a prominent volcanic ridge—it is the underwater continuation of the Hawaiian Islands chain. What might explain the sharp bend in the ridge?

When did the oceans form?

Where did the water come from?

How do we measure ocean depth?

What are the important features of ocean basins?

The Composition of Seawater and the Movement of Sediment

f you have ever swum in the ocean, you'll know that seawater contains a lot of salt, making it not only unpalatable but also dangerous for human consumption. In fact, if you drank only salt water you would become dehydrated because your body would try to rid itself of the excessive amounts of ions, such as sodium and potassium, contained in salt water.

The **salinity**, or salt content, of seawater ranges from 3.3 to 3.7%. This may sound quite small, but it is about 70 times the salinity of tap water. Not surprisingly, when seawater is evaporated, more than three-quarters of the dissolved matter is sodium chloride—that is, table salt. Seawater contains most of the other natural elements as well, but many of them are present in such low concentrations that only extremely sensitive analytical instruments can detect them.

salinity A measure of the salt content of a solution.

The elements dissolved in seawater come from several sources. Chemical weathering of rock releases soluble materials such as salts of sodium, potassium, and sulfur. Having been dissolved out of the rock, the soluble compounds become part of the dissolved load in river water flowing to the sea. Volcanic eruptions, both on land and beneath the sea, also add soluble compounds to water via volcanic gases and hot springs. The *black smoker* vents found at spreading centers add dissolved minerals from the hot rocks in the midocean rift (**FIGURE 12.4**). Two other processes, evaporation of surface water and freezing of seawater, tend to make seawater saltier because they remove fresh water while leaving the salts behind.

Why, then, doesn't the sea become saltier over time? The answer is that it constantly receives fresh water from *precipitation* and *river flow* (**FIGURE 12.5**). Also, aquatic plants and animals withdraw some elements, such as silicon, calcium, and phosphorus, to build their shells or skeletons. Other elements precipitate out in mineral form and settle to the seafloor, and chemical reactions between seawater and hot volcanic rocks remove some salts from solution. All of these processes balance each other, keeping the composition of seawater essentially unchanged. This is the problem that John Joly ran into when he tried to use the salinity of seawater to estimate the age of the ocean (Chapter 10).

Black smoker vents FIGURE 12.4

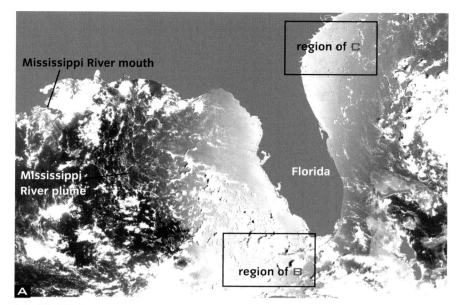

July 30, 2004

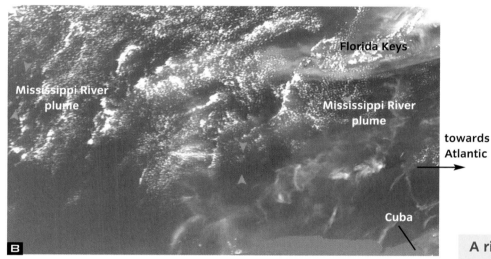

August 4, 2004

August 9, 2004

A river runs through it
FIGURE 12.5

These satellite photos, taken in 2004, show a "river" (or, as Earth scientists call it, a *plume*) of less salty water running from the mouth of the Mississippi River through (A) the Gulf of Mexico and (B) the Florida Straits and into (C) the Atlantic Ocean, before it finally mixes sufficiently with ocean water that it cannot be seen any more. The river water appears brown in the photos. Shipboard measurements showed that the plume was 50 km wide and 10 to 20 m deep. It had lower salinity, higher temperature, and double the chlorophyll concentration of the other water in the Florida Straits. Scientists estimate that 23% of the Mississippi River's discharge flows directly to the Atlantic Ocean in this manner.

TURBIDITY CURRENTS

In 1929, an earthquake occurred in the Grand Banks region off the coast of Newfoundland, triggering a submarine landslide. This generated an underwater current, laden with sediment, and within the next 13 hours, the passing current had snapped a series of transatlantic telephone cables. Scientists knew the positions of the breaks in the cables and the times at which they broke, so they could work out that the current traveled at a speed of about 75 km/h.

To understand what happened, let's look at how seawater transports sediment in more detail. In Figure 12.3, we can see that the shallowly submerged *continental shelves* pass abruptly into *continental slopes*. These slopes descend to depths of several kilometers. Sediment deposited on the edge of the shelf is poised for further transport down the slope and onto the adjacent continental rise.

Marine scientists have found thick bodies of coarse sediment lying at the foot of the continental slope at depths as great as 5 km (**Figure 12.6A**). At first, it was difficult to explain where these coarse accumulations came from, but eventually scientists demonstrated that the sediments could be deposited by **turbidity currents**. These are gravity-driven currents that contain dilute mixtures of sediment and water with a density greater than that of the surrounding water. Similar currents have been reproduced in the laboratory, using dense mixtures of water, silt, and clay in water-filled tanks. They have also been discovered moving across the floors of lakes and reservoirs (**Figure 12.6B**). In the oceans, turbidity currents are set off by earthquakes, as in the Grand Banks example, or by landslides, and major coastal storms. Off the mouths of rivers, they can be set in motion by large floods.

> **turbidity current**
> A gravity-driven current consisting of a dilute mixture of sediment and water with a density greater than that of the surrounding water.

Deep-sea turbidites Figure 12.6

A These deep-sea turbidite beds have been tilted, uplifted, and exposed in a wave-eroded beach along the coast of the Olympic Peninsula in Washington. ▶

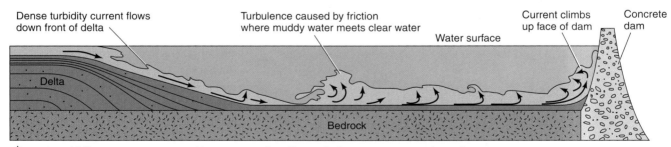

▲ **B** A turbidity current generated by a surge of sediment-laden water enters the quiet water of a reservoir behind a dam. Moving rapidly down the face of a delta, the current passes along the lake floor and climbs up the face of the dam before subsiding. As the sediment settles to the bottom, it forms a graded *turbidite* layer.

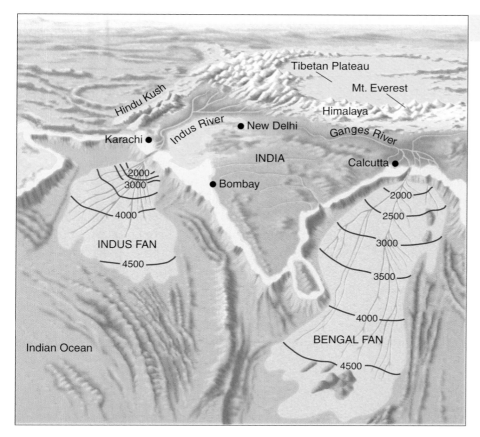

Deep-sea fans FIGURE 12.7

The Indus and Ganges–Brahmaputra river systems have built the vast deep-sea fans on the seafloor adjoining the Indian subcontinent. Most of the sediment in the fans was transported from the high Himalaya to the north as the mountain system was uplifted over tens of millions of years. Contours are water depths in meters.

Turbidity currents are effective geologic agents on continental slopes, where they can reach velocities greater than those of the swiftest streams on land. Some achieve a velocity of more than 90 km/h and transport up to 3 kg/m³ of sediment, spreading it as far as 1000 km from its source.

A turbidity current typically deposits a graded layer of sediment called a **turbidite** (Figure 12.6B). We see

> **turbidite** A graded layer of sediment that is deposited by a turbidity current.

these graded layers because the moving current that formed them was continuously, and rapidly, losing energy. As a rapidly flowing turbidity current slows down, it deposits successively finer sediment. Turbidites are deposited relatively rarely at any one site on the continental rise or an adjacent abyssal plain, perhaps only once every few thousand years. In these places, far distant from the source of the sediment, the deposits are

thin layers a few millimeters to 30 cm thick. Over millions of years, turbidites can slowly build up to form vast deposits that stretch beyond the continental realm.

Deep-sea fans are a dramatic example of just how far land-derived sediment in the ocean can extend beyond the continental shelves. When shelves are exposed at times of lowered sea level and rivers extend across them nearly to the continental slope, the stage is set for the rapid formation of deep-sea fans. Some large submarine canyons on the continental slopes are aligned with the mouths of major rivers like the Amazon, Congo, Ganges, and Indus. At the base of many such canyons is a huge deep-sea fan, a fan-shaped body of sediment that spreads downward and outward to the deep seafloor (FIGURE 12.7). The sediments, which are derived mostly from the land, include fragments of land plants as well as fossils of shallow- and deepwater marine organisms. We also find many graded layers that we can recognize as turbidites.

BIOTIC ZONES AND DEEP-SEA SEDIMENTS

We can classify plants and animals according to which part of the ocean—that is, which biotic zone—they live in (FIGURE 12.8). Plants and animals living in the uppermost waters of the ocean occupy the *pelagic* zone and are called *pelagic organisms*. This includes animals that swim freely under their own locomotion, such as reptiles, squids, fish, and marine mammals. *Benthic organisms* live on the bottom or in bottom sediments (the *benthic* zone). Floating or drifting (*planktic*) organisms include phytoplankton, which are mainly single-celled plants, and zooplankton, which are tiny animals. Among the most important zooplankton are single-celled foraminifera and radiolaria. Foraminifera have a calcareous shell and, as we will see in Chapter 16, they play an important role in helping Earth scientists identify changes in climate over the years. Radiolarian remains, by contrast, consist of silica.

Because plant life needs enough light energy for *photosynthesis*, it is restricted to the upper 200 m of the ocean (the *photic* zone). Plants also require nutrients, which are available mainly along coasts and shallow continental margins, and as a result plant life is limited over much of the deep ocean.

By analyzing samples from *sediment cores*, Earth scientists can work out the sources of seafloor sediment (FIGURE 12.9). (We will describe sediment cores in detail in Chapter 16.) A large portion of this sediment is produced by biological activity in the surface waters, but some sediment is transported over great distances from continental interiors, especially by winds, and eventually reaches the deep sea (see Figure 1.3).

The deep seafloor is mantled with the skeletal remains of single-celled planktonic and benthic animals and plants, which cover vast areas (look ahead to Figure 16.3A). When more than 30% of the surface sediment is made up of such remains, we call it a *calcareous ooze* or *siliceous ooze*, depending on the chemical compo-

Marine organisms FIGURE 12.8

A A leopard seal bares teeth in a threat display to protect her kill, Antarctic Peninsula, Antarctica.

B This tube anemone, waving its tentacles, is a benthic organism.

C Antarctic krill (*Euphausia superb*), a small shrimp-like crustacean, is the most important zooplankton in the Antarctic food web, Weddell Sea Antarctica. The yellow color comes from algae in its stomach.

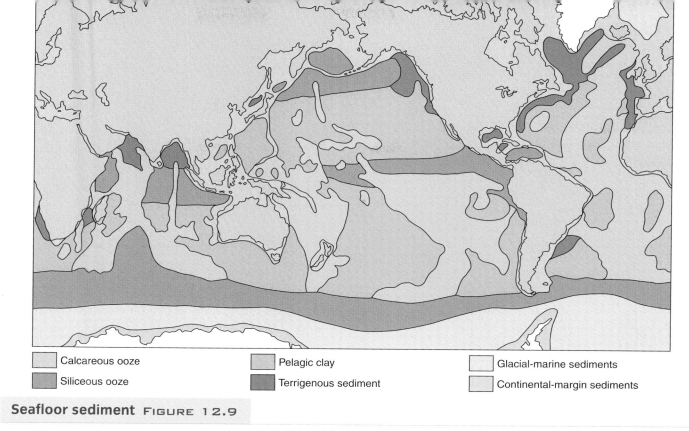

| Calcareous ooze | Pelagic clay | Glacial-marine sediments |
| Siliceous ooze | Terrigenous sediment | Continental-margin sediments |

Seafloor sediment FIGURE 12.9

Map of the World Ocean showing the distribution of the principal kinds of sediment on the ocean floor.

sition of its major component. Calcareous ooze covers broad areas of low to middle latitudes, because the warm surface waters in those latitudes favor the growth of carbonate-secreting organisms (Figure 12.9). Their tiny shells settle to the seafloor in vast numbers, but they accumulate at an average rate of only about 1–3 cm per thousand years.

We do not find calcareous ooze in regions where the water is unusually deep, even at low to middle latitudes. That's because the cold, deep ocean water is under high pressure and contains more dissolved carbon dioxide than is contained in shallower waters, making the waters more acidic. For this reason, these deep waters can readily dissolve any carbonate particles that reach their level. In the Pacific Ocean, this level lies at 4000–5000 m, whereas in the Atlantic it is somewhat shallower. This explains why calcareous ooze is missing over large portions of the deep north and south Pacific Ocean and some marginal parts of the Atlantic Ocean.

Siliceous organisms make up the major component of the bottom sediments in broad belts across the equatorial and far northern Pacific Ocean, sectors of the Indian Ocean, and a belt around the Southern Ocean.

Sediment is composed of *lithic* clasts of rock fragments from the continents' mantles—the continental shelves and slopes. These sediments are also found at locations where debris-laden glaciers generate icebergs that raft sediment seaward from glacier margins (see Chapter 6) and where turbidity currents have transferred sediment from the shelves to the deep-sea floor. Finally, far from land in regions of low biological activity, we find very fine-grained reddish or brownish clay, generally called *red clay*. Much of the clay is made up of fine windblown dust. The dust is red because it contains oxidized iron-rich minerals.

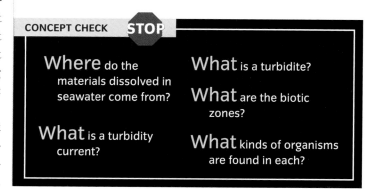

CONCEPT CHECK **STOP**

Where do the materials dissolved in seawater come from?

What is a turbidity current?

What is a turbidite?

What are the biotic zones?

What kinds of organisms are found in each?

Ocean Water and Its Circulation

LEARNING OBJECTIVES

Identify the layers of the ocean.

Describe ocean currents and why they occur.

Explain the changes in Pacific Ocean currents that cause El Niño and La Niña.

Discuss the ocean conveyor belt.

Explain how the oceans regulate climate.

The salinity and temperature of seawater combine to control its density: Cold, salty water is dense and will sink, whereas warm, less salty water is less dense and will rise. Looking back at Figure 12.5, we can see this very clearly: The warm and less salty (low-density) water from the Mississippi River remains on top of the water in the Gulf of Mexico for hundreds of kilometers.

Ocean scientists have discovered three major layers, or zones, in the ocean, in each of which the density of the water differs. These are the surface layer, the transi-

tional zone (the *thermocline*), and the deep zone. The differences are caused by changes in both temperature and salinity, with temperature being the major factor (**FIGURE 12.10**). Warm water is less dense than cold water, and saltier water is denser than less salty water. Once established, the pressure differences induce the water to flow.

OCEAN CURRENTS

In 1492, when Christopher Columbus set sail across the Atlantic Ocean in search of China, he took an indirect route. On the outward voyage, he sailed southwest toward the Canary Islands and then west on a course that carried him to the Caribbean Islands, where he first sighted land. In choosing this course, he was taking advantage of favorable winds and surface **ocean currents**, rather than fighting the westerly winds and currents that would have hindered his progress if he had taken a more

> **ocean current** A persistent, dominantly horizontal flow of water in the ocean.

Ocean layers FIGURE 12.10

The three major density zones in the ocean: the surface layer, the transitional zone (*thermocline*), and the deep zone.

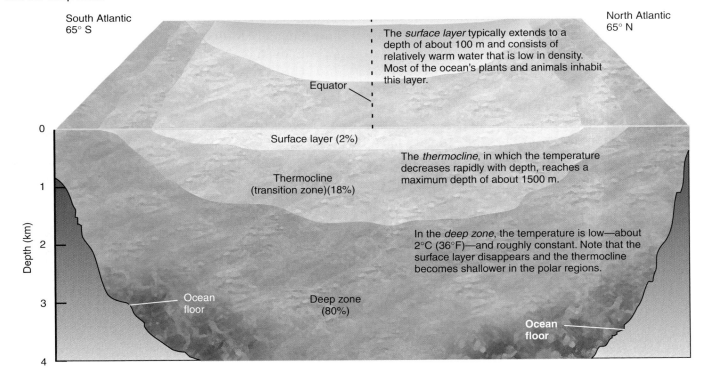

South Atlantic 65° S

North Atlantic 65° N

Equator

The *surface layer* typically extends to a depth of about 100 m and consists of relatively warm water that is low in density. Most of the ocean's plants and animals inhabit this layer.

Surface layer (2%)

Thermocline (transition zone)(18%)

The *thermocline*, in which the temperature decreases rapidly with depth, reaches a maximum depth of about 1500 m.

Depth (km)

In the *deep zone*, the temperature is low—about 2°C (36°F)—and roughly constant. Note that the surface layer disappears and the thermocline becomes shallower in the polar regions.

Ocean floor

Deep zone (80%)

Ocean floor

SEA-SURFACE TEMPERATURE

Because the heat-holding capacity of water is so high, sea-surface temperature has an important effect on both weather and climate. Slow changes in ocean–atmosphere systems, such as El Niño, the North Atlantic Oscillation, and the Pacific Decadal Oscillation, can be felt in climate cycles over many years.

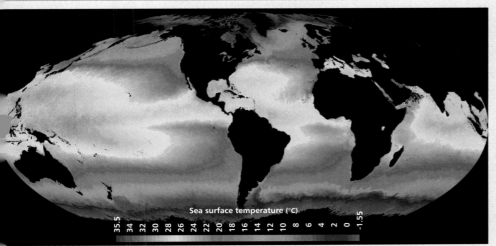

Sea surface temperature (°C)

35.5 34 32 30 28 26 24 22 20 18 16 14 12 10 8 6 4 2 0 -1.55

◄ **Sea-surface temperature**
While ocean temperatures are generally warm near the Equator and cold at high latitudes, winds and currents distort this simple picture. Thus, the western equatorial Pacific is warmer than the eastern half, and warm equatorial currents move poleward in the form of major ocean currents such as the Gulf Stream.

▼ **El Niño–La Niña**
El Niño is the appearance of warm water in the eastern equatorial Pacific (left) near Christmas time. El Niño is often followed by La Niña, characterized by cold ocean temperatures across the equatorial Pacific. These local ocean properties are connected to global changes in atmospheric patterns.

▼ **North Atlantic Oscillation**
The weather of northern Europe has dramatic swings every 5–10 years in connection with the North Atlantic Oscillation. When the air pressure over Portugal becomes larger than that over Iceland, the westerly wind in the Atlantic becomes stronger and brings warm and wet marine air to northern Europe for a mild winter. The opposite brings a colder and drier winter. The extreme warm or cool phases can last for as long as two decades. Ocean temperatures suggest a reversal to cool Pacific Decadal Oscillation conditions in 1998.

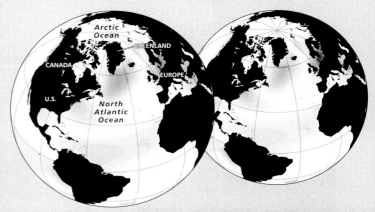

heat and maintains the temperature of the water around it. Both of these processes tend to reduce the amount of variation in ocean temperatures (**Figure 12.15**). The ocean in turn moderates the climate of coastal regions on land. Along the Pacific coast of Washington and British Columbia, for example, winter air temperatures seldom drop to freezing, whereas inland temperatures plunge to −30°C or lower.

The thermohaline portion of the ocean's global conveyor belt is one of the most important natural influences on climate. Because it redistributes heat from the tropics to the poles, any weakening of the circulation would be expected to increase the temperature contrast between the two. The cause and effect may work the other way, too: During the last ice age, the polar ice caps may have cut off the "conveyor belt," which would have made the glaciation longer-lasting and more severe. In Chapter 16, we will look at the possibility of rapid climate change triggered by a shutdown of the thermohaline circulation.

The global air temperature distribution FIGURE 12.15

These maps show the average daily air temperature in (A) January and (B) July over both land and sea. Notice the much broader range of temperatures over land in both maps. Also, note that there is a pronounced seasonal change in land temperatures, whereas in the world's oceans there is little difference between summer and winter (especially in the tropics).

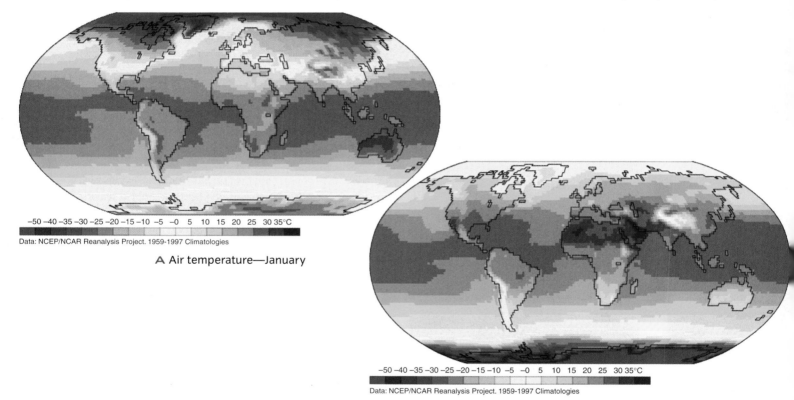

−50 −40 −35 −30 −25 −20 −15 −10 −5 −0 5 10 15 20 25 30 35°C

Data: NCEP/NCAR Reanalysis Project. 1959-1997 Climatologies

A Air temperature—January

−50 −40 −35 −30 −25 −20 −15 −10 −5 −0 5 10 15 20 25 30 35°C

Data: NCEP/NCAR Reanalysis Project. 1959-1997 Climatologies

B Air temperature—July *Data: NCEP/NCAR Reanalysis Project; 1959–1997, Climatologies.*

CONCEPT CHECK **STOP**

What are the three layers of the ocean?

What drives surface ocean currents?

How are pressure differences created in ocean water?

How do El Niño and La Niña conditions affect Peruvian coastal upwelling and trade winds?

What is the ocean conveyor belt?

Amazing Places: Monterey Bay, California

Just offshore from the Monterey Peninsula, California, the Monterey Submarine Canyon bisects Monterey Bay, plunging to about 1.8 km, making it comparable in depth to the Grand Canyon (A). The near-shore presence of a deepwater canyon provides a coldwater upwelling—produced by offshore winds—that is rich in nutrients that can support unusually abundant food for seabirds, whales, dolphins, and sea otters (B). The floor of Monterey Bay is host to a number of other weird creatures. This three-inch long nudibranch (C) is called the sea shawl. The nudibranch is a carnivorous organism, with some species even eating other nudibranchs. A rosy rockfish swims by a basket starfish (D).

Global Locator

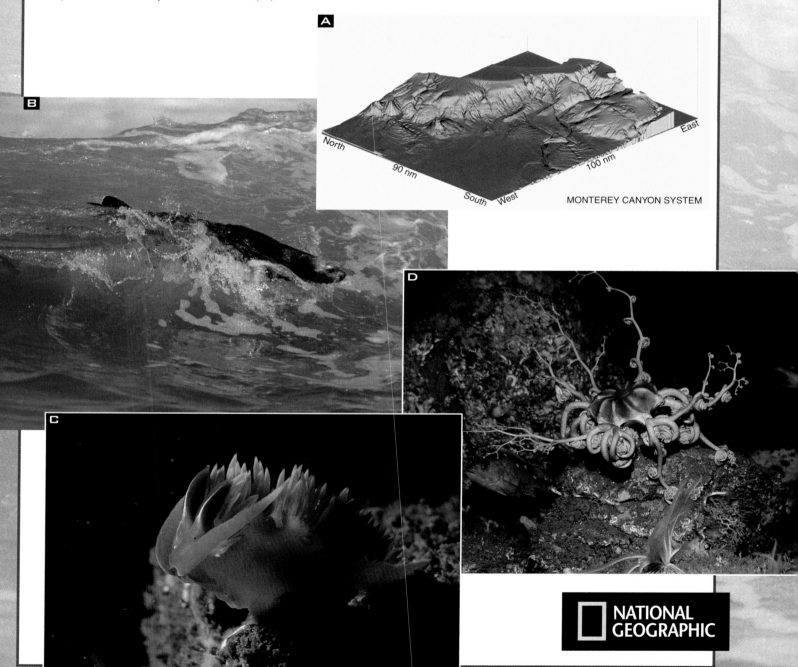

SUMMARY

1 The Ocean Basins

1. There has been an ocean on Earth for at least 4 billion years.

2. Most of the ocean water is contained in four huge, interconnected **basins**—the Pacific, Atlantic, Indian, and Southern Oceans.

3. The water in the ocean may have condensed from steam produced by primordial volcanic eruptions, been delivered to the planet's surface via cometary impacts, or both.

4. Ocean basins are marked by a *mid-oceanic ridge* with a *central axial rift* and by *abyssal plains*.

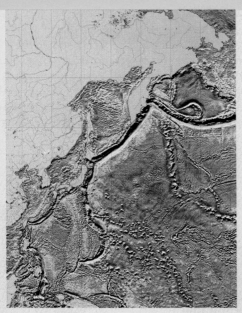

2 The Composition of Seawater and the Movement of Sediment

1. Seawater ranges in **salinity** from 3.3 to 3.7%. Freezing and evaporation make the water saltier, whereas rain, snow, and *river flow* make it less salty.

2. **Turbidity currents** have built thick deposits of sediment at the base of the *continental slope* and on the adjacent abyssal plain.

3. The chief kinds of sediment on the deep seafloor are brownish or reddish clay (blown in from continents by the global wind patterns), calcareous ooze, and siliceous ooze. The distribution of oozes is related to surface-water temperature and water depth.

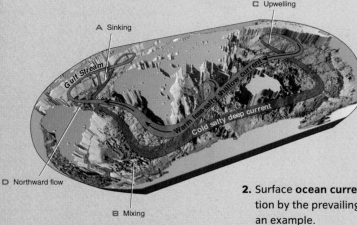

3 Ocean Water and Its Circulation

1. Ocean water forms layers based on density, which is controlled by temperature and salinity. These layers are the surface layer, the thermocline, and the deep layer.

2. Surface **ocean currents** are set in motion by the prevailing winds. **El Niño** is an example.

3. The **thermohaline circulation** involves deep currents that are powered by changes in temperature and salinity in surface waters, which change the water's density and thereby cause it to rise or sink. It is also known as the *ocean conveyor belt*. The thermohaline circulation helps regulate climate.

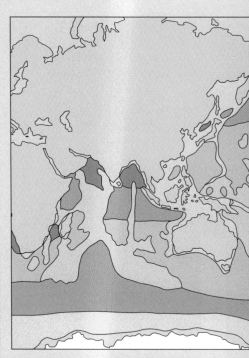

KEY TERMS

1. What kind of evidence might be sought to establish the existence of an ocean early in Earth's history?

2. Refer back to Figure 1.11 and look at the distribution of land and oceans 200 million years ago. How might surface ocean currents have flowed in the world ocean at the time? What would the climate have been in the center of the large landmass (Pangaea)?

3. Some scientists have proposed controlling climate change by increasing the photosynthetic productivity of microscopic marine algae in the oceans. How could marine algae influence the climate? How could their productivity be increased? Do you think that this is a realistic way to counter climate change?

4. Earth scientists have proposed that the eruptions of the Indonesian volcanoes Krakatoa in 1883 and Tambora in 1815 had lasting effects on the behavior of the oceans, leading to a slight drop in sea level. How could a volcanic eruption have led to this effect?

5. During the early Tertiary Period, North and South America were not connected. How might that have changed the global pattern of ocean circulation? Do some research and suggest some ways of testing your hypothesis.

What is happening in this picture ?

This striking image, acquired by a NASA satellite, shows land and sea surface temperatures for a week in April. The eastern coast of North America is on the left and the Atlantic Ocean is on the right. The color scale ranges from deep red (hot) to violet (cold). The Gulf Stream, which carries warm water northward, stands out as a tongue of red and yellow extending from the Caribbean along the coasts of Florida and the Carolinas.

Can you pick out the tip of the Labrador current, which carries cold water southward?

Looking at the land, note the purple colors of the northern Great Lakes. What might this coloration indicate?

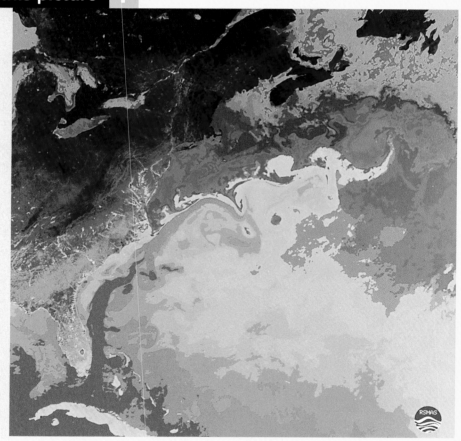

SELF-TEST

1. The ocean formed between _____ billion and _____ billion years ago.
   - a. 3.0/3.56
   - b. 3.56/4.0
   - c. 4.0/4.56
   - d. 4.56/5.0

2. Today, scientists measure the depth of oceans with _____.
   - a. acoustical instruments
   - b. weighted wires
   - c. laser imaging
   - d. optical instruments

3. Much of the oceanic crust is less than _____ years old, while the bulk of the continental crust is over _____ years old.
   - a. 50 million/5 billion
   - b. 100 million/4 billion
   - c. 40 million//1 billion
   - d. 60 million/1 billion

4. The diagram shows a _____ plate boundary.
   - a. collision
   - b. spreading
   - c. subduction
   - d. transform

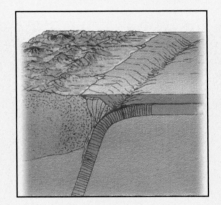

5. Some continental margins are _____ and accumulate thick deposits of continental sediments, while other continental margins are _____ and have trenches marking the location at which ocean crust is sliding beneath continental crust.
   - a. passive/active
   - b. passive/tectonic
   - c. active/passive
   - d. submerging/emerging

6. The salinity of seawater is roughly _____ times the salinity of tapwater.
   - a. 0.7
   - b. 7
   - c. 70
   - d. 700

7. Which of the following processes increases the salinity of seawater?
   - a. freezing
   - b. rain, snow, and river runoff
   - c. evaporation
   - d. Both a and c are correct.

8. Which of the following processes decreases the salinity of seawater?
   - a. freezing
   - b. rain, snow, and river runoff
   - c. evaporation
   - d. Both a and c are correct.

9. Turbidity currents (above) are driven by _____.
   - a. prevailing winds
   - b. gravity
   - c. the Coriolis effect
   - d. temperature differences

Shoreline protection FIGURE 13.20

A Breakwaters Breakwaters constructed along the shore at Tel Aviv in Israel protect the beach from the onslaught of waves, but they have turned a straight coastline into a scalloped one. Wave action around the barriers has added sediment to the beach behind each breakwater, producing a scalloped coastline.

Breakwaters

B Groins Severe storms during the winter of 1993 carved out an inlet in this barrier beach on the south shore of Long Island, New York. A system of groins has trapped sand, protecting the far stretch of beach. The beach in the foreground, without groins, has receded well inland of the houses that were once located on its edge.

Groins

groin, where the beach sand is not being replenished. The net effect, once again, is to protect one part of a beach at the expense of another part.

Another way to protect an eroding beach is to bring in sand and pile it on the beach at the updrift end. Surf then erodes the pile and drifts the new sand down the length of the beach. As you can imagine, constantly feeding a beach with sand in this way can be expensive.

EFFECTS OF HUMAN INTERFERENCE

Many beaches around the world are deteriorating because of human interference. In Southern California, for example, most of the sand on beaches is supplied not by erosion of wave-cut cliffs but by alluvium carried to the sea at times of flood. But inhabitants have built dams across the streams to protect buildings and other struc-

CASE STUDY

The Black Sea Coast

A dramatic example of human interference can be seen along the Russian coast of the Black Sea (A). Ninety percent of the sand and pebbles that form the natural beaches there used to be supplied by rivers as they entered the sea. During the 1940s and 1950s, three things happened: Large resort developments were built at the

beaches (B); large breakwaters were constructed so that two major harbors could be extended into the sea; and dams were built across some rivers inland from the coast (C).

All this construction upset the balance among the supply of sediment to the coast, longshore currents and beach drift, and the deposition of sediment on beaches. By 1960, the combined area of all beaches along the coast had decreased by half. Then beachfront buildings began to sag or collapse as the surf ate away at their foundations. Ironically, many of the resort buildings had been constructed from a concrete aggregate made from large volumes of sand and gravel that had been removed from the beaches. The same concrete had also been used to build the dams that cut off the supply of sediment to the coast.

A

1. When the height between high and low tide is large, as in the Bay of Fundy, engineers sometimes consider using the tide as a way to generate electricity. What might some of the negative consequences be from the installation of tidal power plants?

2. If you live or go to school near a coastline, identify the kind of coastline that is near. How will the rise of a meter in sea level affect your area?

3. Scientists who study climate predict that as the climate gets warmer, severe storms and particularly hurricanes, will become stronger and more frequent. Which regions of North America are most vulnerable and likely to be affected by the changes?

4. When building a house, it is important to assess potential hazards. What advice would you offer to someone planning to build a shoreline home in the following places: (a) southern Georgia; (b) Oregon, south of the Columbia River; (c) the Yucatan Peninsula, Mexico; (d) the south side of Nantucket Island, Massachusetts.

5. Tsunamis have been recorded in the Atlantic Ocean, but they are much less common than in the Pacific and Indian Oceans. Why is this so?

What is happening in this picture ?

This photo shows the aftermath of Hurricane Dennis, in 1999, on the barrier beach of North Carolina's Outer Banks. A battered house, suspended on stilts, sits in the surf on a sandy Atlantic beach. Its windows and doors are sealed with plywood, and waves break underneath it. A pile of wooden debris is heaped up nearby.

◼ What might have happened here?

1. Which of the following statements about global sea level is true?
   a. Global sea level is constant over both human and extended time periods.
   b. Global sea level can vary by tens to hundreds of meters over extended periods.
   c. Global sea level varies by only a few meters over extended periods.
   d. Large variations in global sea level can be an effect of climate change.
   e. Both (b) and (d) are correct.

2. This illustration depicts the forces involved in the generation of tides on Earth. Label the illustration with the following terms:
   a. distorted water level
   b. common center of mass of Earth–Moon system
   c. inertial force
   d. uniform level of water (theoretical)
   e. gravitational force
   f. tide-raising force

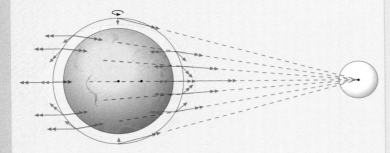

3. The most important agent shaping coastal landforms is the action of _____.
   a. storms
   b. streams
   c. salinization
   d. waves

4. A _____ is a localized narrow channel of water that can pull unwary swimmers out to sea.
   a. rip current
   b. surf zone
   c. breaking current
   d. tidal run-up

5. When a wave reaches the shore it breaks, expending most of its energy. Label the following features or regions on the diagram: (a) breaking wave, (b) surf, (c) turbulent water region, (d) the region where the wave first "feels bottom."

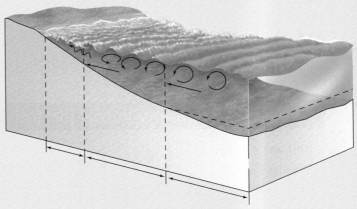

6. Littoral drift, the process through which sediment is transported along a beach, includes _____ and _____.
   a. beach drift/ebb tide
   b. ebb tide/longshore drift
   c. flood tide/ebb tide
   d. longshore current/beach drift

7. Tidal currents are made up of two opposing currents called _____ currents.
   a. longshore and littoral
   b. ebb and flood
   c. longshore and flood
   d. ebb and littoral

8. The most common type of coast is the _____.
   a. barrier island coast
   b. coral-reef coast
   c. lowland beach
   d. rocky coast

9. This diagram shows some of the landforms created by wave action against a marine cliff. Label the following features on this diagram: (a) arch, (b) cave, (c) notch, (d) stack.

10. Broad expanses of isolated shallow water called _____ are common features immediately adjacent to barrier-island coastlines.
    a. salt marshes
    b. marine terraces
    c. lagoons
    d. tidal inlets

11. A _____ is a ridge of sand or gravel that joins a barrier island to the mainland.
    a. spit
    b. tombolo
    c. bay barrier
    d. lagoon

12. Coral-reef coasts are built up of _____ secreted by organisms as their skeletal material.
    a. calcium carbide
    b. calcium carbonate
    c. calcium chloride
    d. calcium chlorate

13. Which of the following events could not trigger a tsunami?
    a. hurricane
    b. earthquake
    c. landslide
    d. volcanic eruption

14. _____ is often used to protect beaches, as shown in this photo of a barrier beach on the south shore of Long Island, New York.
    a. Artificial nourishment
    b. A system of breakwaters
    c. A system of groins
    d. A seawall

15. Which of the following factors did not contribute to the deterioration of the Russian coast of the Black Sea during the 1940s and 1950s?
    a. construction of beach resorts
    b. construction of breakwaters
    c. construction of dams
    d. construction of groins

The Atmosphere: Composition, Structure, and Clouds

14

Around 65 million years ago, the last of the dinosaurs died out. Today, we hypothesize that the extinction was probably triggered by a great meteorite impact, but it's less well known that dinosaur numbers were declining long before the meteorite struck. Why was this? One hypothesis is that the dinosaurs were suffocated by subtle changes in our planet's atmosphere (here viewed from above by astronauts).

Like us, the dinosaurs required oxygen to breathe. But the oxygen content of the atmosphere has varied through Earth's long history. By exploring several lines of evidence, especially by analyzing samples of air trapped in amber (a fossil tree resin), scientists discovered that 100 million years ago, when the dinosaurs were in their prime, the oxygen content of the atmosphere was 40% higher than it is today. Thanks to an oxygen-rich atmosphere, the dinosaurs needed only relatively small lungs to breathe.

However, near the end of the dinosaurs' reign, the oxygen level was declining. It is hypothesized that dinosaurs, with large bodies and small lungs, were unable to adjust to the decline.

Humans are less vulnerable to such changes—we can survive with oxygen levels ranging from 40% above to 44% below those in today's atmosphere. If we could turn the clock back, we could breathe the same air as the dinosaurs. However, we shouldn't be too complacent. The balance of gases in our atmosphere also protects us from harmful solar rays and plays a crucial role in warming our planet and transporting water around the globe. We must be careful not to upset this balance.

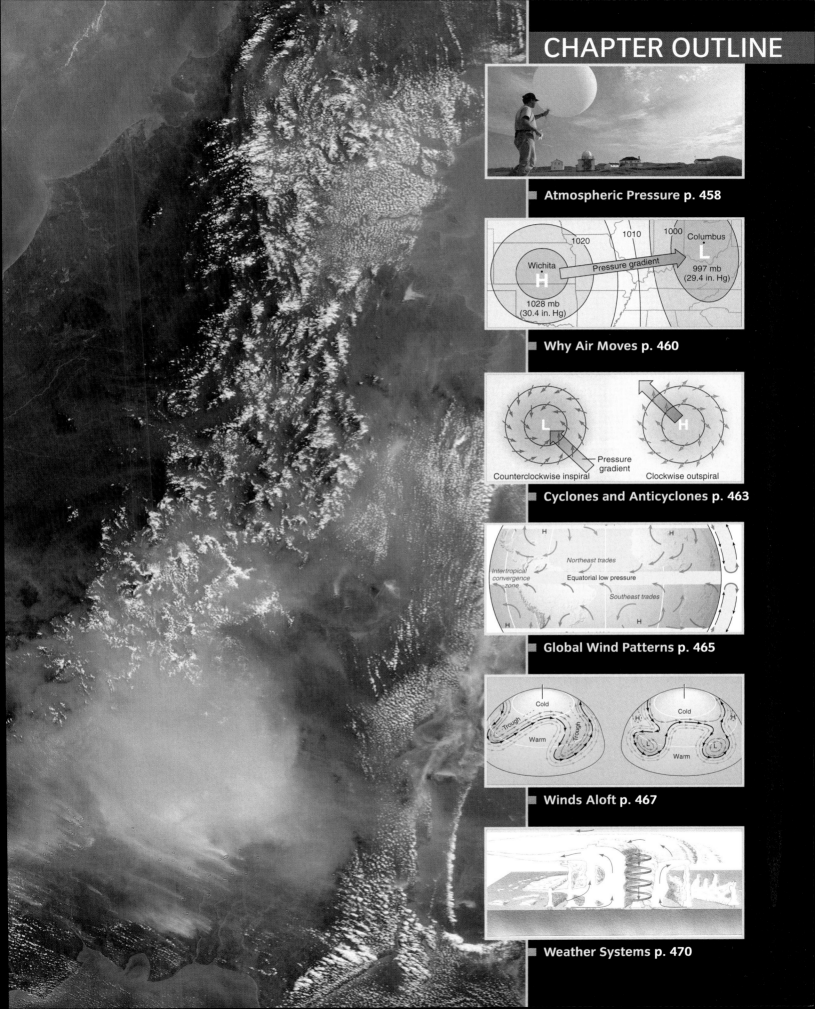

Atmospheric Pressure

LEARNING OBJECTIVES

Define atmospheric pressure.

Explain how a barometer works.

Discuss how air pressure varies with altitude.

W e live at the bottom of a vast ocean of air—Earth's *atmosphere* (FIGURE 15.1). Like the water in the ocean, the air in the atmosphere is constantly pressing on Earth's surface and on objects on the surface.

The atmosphere exerts pressure because *gravity* pulls the gas molecules in the air toward Earth. Gravity is a force of attraction that acts between all masses—in this case, between gas molecules and Earth's vast bulk.

Atmospheric pressure is produced by the weight of a column of air above a specified area of Earth's surface.

When TV weather forecasters talk about "highs" and "lows," they are referring to air pressure that is higher or lower than average. At sea level, about 1 kilogram of air presses down on each square centimeter of surface (1 kg/cm^2)—about 15 pounds on each square inch of surface (15 $lb/in.^2$).

The basic metric unit of pressure is the *pascal* (Pa). Pressure is measured in terms of the force (measured in units called *newtons* [N]) bearing down on a certain area, and 1 Pa = 1 N per m^2. At sea level, the average pressure of air is 101,320 Pa. Many atmospheric pressure measurements are reported in *bars* and *millibars* (mb) (1 bar = 1000 mb = 10,000 Pa). In this book, we will use the millibar as the metric unit of atmospheric pressure. Standard sea-level atmospheric pressure is 1013.2 mb.

atmospheric pressure Pressure exerted by the atmosphere at Earth's surface because of the force of gravity acting on the overlying column of air.

An ocean of air FIGURE 15.1

Earth's terrestrial inhabitants live at the bottom of an ocean of air. This buoyant weather balloon, known as a *radiosonde*, is on Sable Island, Nova Scotia. It carries instruments that measure temperature and pressure at high atmospheric levels and radios the data to scientists on the ground.

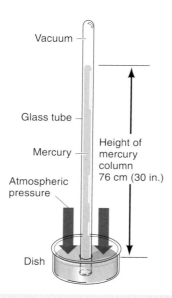

The mercury barometer FIGURE 15.2

Atmospheric pressure pushes the mercury upward into the tube, balancing the pressure exerted by the weight of the mercury column. As atmospheric pressure changes, the level of mercury in the tube rises and falls.

You probably know that a **barometer** measures atmospheric pressure. But do you know how it works? It's based on the same principle as drinking soda through a straw. When using a straw, you create a partial vacuum in your mouth by lowering your jaw and moving your tongue. The pressure of the atmosphere then forces soda up through the straw. A similar process occurs in a barometer (FIGURE 15.2).

barometer An instrument that measures atmospheric pressure.

AIR PRESSURE AND ALTITUDE

If you have felt your ears "pop" during an elevator ride in a tall building or on an airplane that is climbing or descending, you've experienced a change in air pressure related to altitude.

In 1658, a young French scientist, Blaise Pascal (1623–1662), carried out an important experiment to prove that air pressure decreases with altitude. He arranged for rock climbers to ascend a prominent volcanic rock known as Puy-de-Dôme, measuring the air pressure at several places during the ascent. Despite the inconvenience of carrying a meter-long glass tube and a flask of mercury up the steep slope of the Puy, the climbers performed the task successfully, and their measurements demonstrated that air pressure decreases with altitude (FIGURE 15.3).

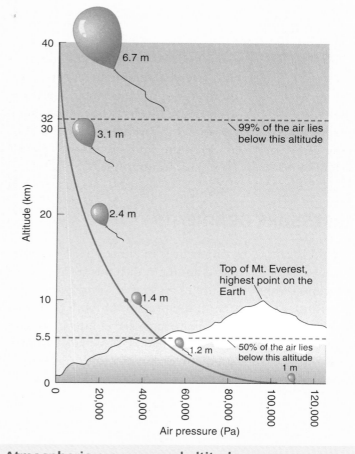

Atmospheric pressure and altitude FIGURE 15.3

Air pressure decreases steadily with altitude. If a helium balloon 1 m in diameter is released at sea level, it expands as it floats upward because of the decrease in pressure. If the balloon did not burst, it would be 6.7 m in diameter at a height of 40 km.

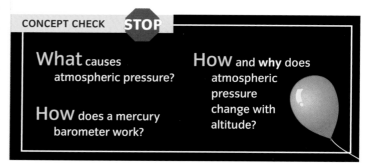

CONCEPT CHECK **STOP**

What causes atmospheric pressure?

How does a mercury barometer work?

How and **why** does atmospheric pressure change with altitude?

Why Air Moves

LEARNING OBJECTIVES

Describe how pressure gradients drive wind.

Discuss local wind systems.

Explain convection loops.

ind is defined as air moving horizontally over Earth's surface. Air movements can also be vertical, but these are referred to by other terms, such as *updrafts* or *downdrafts*. Wind direction is identified by the direction from which the wind comes—a westerly wind blows *from* west *to* east, for example.

PRESSURE GRADIENTS

Wind is caused by differences in atmospheric pressure between one place and another. Air tends to move from regions of high pressure to regions of low pressure, and continues to do so until the pressure in both regions is uniform. On a weather map, lines that connect locations with equal pressure are called **isobars**. A change of pressure, or **pressure gradient**, occurs at a right angle to the isobars (**FIGURE 15.4**).

Pressure gradients develop because of unequal heating in the atmosphere. We can see how this occurs by examining the development of **convection loops**, as shown in **FIGURE 15.5**.

LOCAL WINDS

If you're from Southern California, you're probably familiar with the *Santa Ana*, a fierce, searing wind that often drives raging wildfires into foothill communities (**FIGURE 15.6**). In October 2007, fires driven by Santa Ana

---

isobars Lines on a map drawn through all points having the same atmospheric pressure.

pressure gradient A change of atmospheric pressure, measured along a line at right angles to the isobars.

convection loop A circuit of moving fluid, such as air or water, created by unequal heating of the fluid.

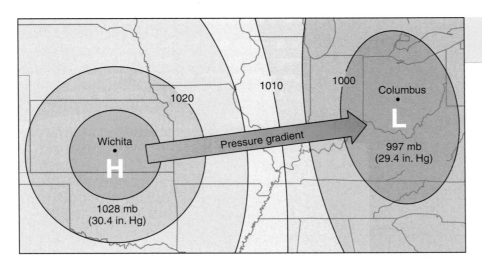

Isobars and a pressure gradient FIGURE 15.4

This figure shows a pressure gradient. Because atmospheric pressure is higher at Wichita than at Columbus, the pressure gradient will push air toward Columbus, producing wind. A greater pressure difference between the two locations would produce a greater force and a stronger wind.

Global Wind Patterns

Describe how Earth–Sun relations influence the development of convection cells.

Explain the role of the Coriolis effect in global wind patterns.

Define the intertropical convergence zone and the trade winds.

Two things energize the atmosphere: the Sun's heat and Earth's rotation. Because Earth is a sphere, the Sun does not warm every place on Earth equally. Only at places where the Sun is directly overhead is the maximum amount of heat received per unit of surface area. That's because at those locations the Sun's rays strike Earth perpendicularly. At all other locations the surface is at an angle to the incoming rays, so they receive less heat per unit of surface area (**FIGURE 15.11**).

As we will see in more detail in Chapter 17, the tilt of Earth's axis also helps determine which points on the surface receive the most heat at different times of the year. For now, it is important to understand that at the spring and fall equinoxes the maximum solar energy (about 1366 watts per square meter) falls on the Equator. In December, because of the 23.5° tilt of Earth's axis of rotation to the plane of its passage around the Sun, solar input is most intense at 23.5° S, the Tropic of Capricorn. In June, the incoming solar energy is most intense at 23.5° N, the Tropic of Cancer (see Figure 17.4 on page 520).

Winds and ocean currents are the natural processes by which the Earth system redistributes the heat more evenly. Currents move heat from the Equator—where the input of solar heat is greatest—toward the poles, where it is least. Unequal heating of Earth's surface causes convection loops in the atmosphere, as we saw in Figure 15.5. Heated air near the Equator expands, becomes lighter, and rises. Near the top of the *troposphere*

Solar intensity and latitude FIGURE 15.11

The angle at which the Sun's rays strike Earth's surface varies from one geographic location to another, owing to Earth's spherical shape and its inclination on its axis.

A Sunlight (represented by the flashlight) that shines vertically near the Equator is concentrated on Earth's surface.

B–C Toward the poles, light hits the surface more and more obliquely, spreading the same amount of radiation over larger and larger areas.

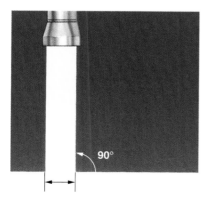

90°

1 unit of surface area

One unit of light is concentrated over 1 unit of surface area.

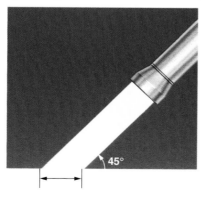

45°

1.4 units of surface area

One unit of light is dispersed over 1.4 units of surface area.

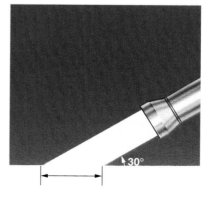

30°

2 units of surface area

One unit of light is dispersed over 2 units of surface area.

Global atmospheric circulation
FIGURE 15.12

Huge convection cells transfer heat from equatorial regions, where the input of solar energy is greatest, toward the poles, where the solar input is least. Because Earth is rotating, the flow of air toward the poles and the return flow toward the Equator are constantly deflected sideways, creating the circulating air masses you may have seen on weather maps. The convection cells are permanent features of Earth's atmosphere and therefore have a great influence both on day-to-day weather and on long-term climate.

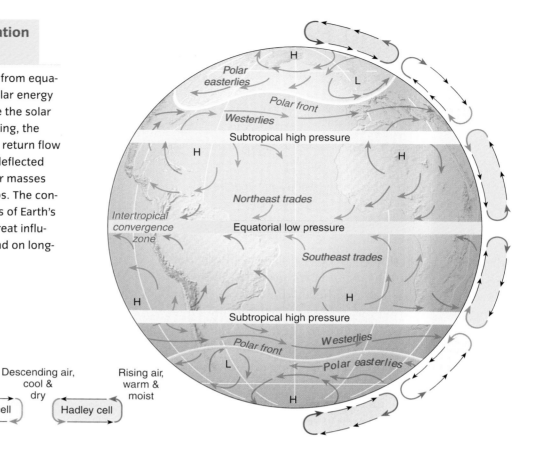

Descending air, cool & dry — Polar cell
Rising air, warm & moist
Descending air, cool & dry — Ferrel cell
Rising air, warm & moist — Hadley cell

it spreads outward toward the poles. As the upper air travels northward and southward toward the poles, it gradually cools, becomes heavier, and sinks. Upon reaching the surface, this cool air flows back toward the Equator, warms up, and rises again, thereby completing a convective cycle. In reality, it is not quite so simple: Global atmospheric circulation actually organizes itself into three *convection cells* that interlock like gears (FIGURE 15.12). We met these same convection cells in Chapter 6 (see Figure 6.1), because they play a major role in the locations of deserts.

WIND SYSTEMS

The Coriolis effect further complicates the picture, breaking up the flow of convective air between the Equator and the poles into belts, as shown in Figure 15.12. For example, a large belt or cell of circulating air lies between the Equator (0°) and about 30° latitude in both the Northern and Southern Hemispheres. Warm air rises near the Equator, creating a low-pressure zone called the **intertropical convergence zone**. The air rises to the top of the troposphere and begins to flow toward the poles, but it veers off course as a result of the Coriolis effect. By the time it reaches a latitude of 30°, the high-altitude air mass has cooled and started to sink. The descending air flows back across Earth's surface toward the Equator. As it flows, the land and sea warm the air so that it eventually becomes warm enough to rise again.

The low-latitude convection cells (from 0° to 30° N and S) created by this circulation pattern are called *Hadley cells*. The prevailing surface winds in Hadley cells are deflected by the Coriolis effect so that they blow northeasterly in the Northern Hemisphere (that is,

intertropical convergence zone (ITCZ) A zone of convergence of air masses along the equatorial trough.

they flow from the northeast toward the southwest), whereas in the Southern Hemisphere they are south-easterly. These winds are called **trade winds** because their consistent direction and flow carried trading ships across the tropical oceans at a time when winds were the chief source of power.

A second set of convecting air cells, called *polar cells*, lies over the polar regions. In a polar cell, frigid air flows across the surface away from the pole and toward the Equator, slowly warming as it moves. When the polar air has reached about latitude 60° north or south, it has warmed enough to rise convectively high into the troposphere and flow back toward the pole, where it cools and descends again, thereby completing the convection cell. Because of the Coriolis effect, the cold air that flows away from the poles is deflected to the right, giving rise to a wind system called the *polar easterlies*.

Between the Hadley cells and the polar cells lies a third, less well-defined set of convection cells. Between about latitude 30° and latitude 60° north and south, we find midlatitude cells, called *Ferrel cells*, in which air flows toward the north and therefore is deflected to

the right by the Coriolis effect, creating a wind that blows from the west. That is why weather systems in the continental United States (and southern Canada) travel from west to east. If you take some time to study Figure 15.12, these circulation patterns will become clearer.

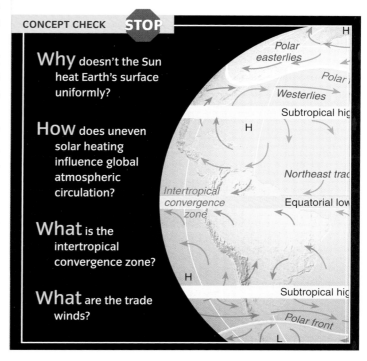

CONCEPT CHECK   STOP

Why doesn't the Sun heat Earth's surface uniformly?

How does uneven solar heating influence global atmospheric circulation?

What is the intertropical convergence zone?

What are the trade winds?

Winds Aloft

LEARNING OBJECTIVES

Explain how pressure gradients develop at upper atmospheric levels.

Discuss the geostrophic wind.

Show how Rossby waves develop and grow.

Define jet streams.

We've looked at air flows at or near Earth's surface, including both local and global wind patterns. But how does air move at the higher levels of the troposphere? As with air near the surface, winds at upper levels of the atmosphere move in response to pressure gradients and are influenced by the Coriolis effect.

A simple physical principle states that pressure decreases less rapidly with height in warmer air than in colder air. Also recall that the solar energy reaching Earth is greatest near the Equator and least near the poles, resulting in a temperature gradient from the Equator to the poles. This gives rise to a pressure gradient; because the atmosphere is warmer near the Equator than the poles, a pressure-gradient force pushes air toward the poles.

THE GEOSTROPHIC WIND

How does a pressure-gradient force pushing toward the poles produce wind, and what will the wind's direction be? Any wind motion is subject to the Coriolis force, which turns it to the right in the Northern Hemisphere and to the left in the Southern Hemisphere. So poleward air motion is toward the east, creating westerly winds in both hemispheres.

Unlike air moving close to the surface, an upper air parcel moves without encountering friction because it is so far from the source of friction—the surface. So there are only two forces operating on the air parcel: the pressure-gradient force and the Coriolis force.

Imagine a parcel of air, as shown in FIGURE 15.13. The air parcel begins to move poleward in response to the pressure-gradient force. As it accelerates, the Coriolis force pulls it increasingly toward the right. As its velocity increases, the parcel turns increasingly rightward until the Coriolis force just balances the gradient force. At that point, the sum of forces on the parcel is zero, so its speed and direction remain constant. We call this type of air flow the **geostrophic wind**. It occurs at upper levels in the atmosphere, and we can see from the diagram that it flows parallel to the isobars.

> ■ **geostrophic wind**
> Wind at high levels above Earth's surface blowing parallel with a system of straight, parallel isobars.

GLOBAL CIRCULATION AT UPPER LEVELS

Figure 15.12 sketches the general air-flow pattern at higher levels in the troposphere. It has four major features: weak equatorial easterlies, tropical high-pressure belts, upper-air westerlies, and a polar low. We've seen that the general temperature gradient from the tropics to the poles creates a pressure-gradient force that generates westerly winds in the upper atmosphere. These *upper-air westerlies* blow in a complete circuit about the Earth, from about 25° latitude almost to the poles. So the overall picture of upper-air wind patterns is quite simple—a band of weak easterly winds in the equatorial

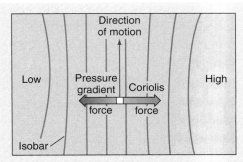

A At upper levels in the atmosphere, a parcel of air is subjected to a pressure-gradient force and a Coriolis force.

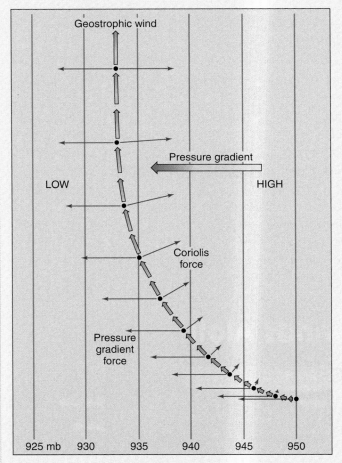

B The parcel of air moves in response to a pressure gradient. At the same time, it is turned progressively sideways until the pressure-gradient force and the Coriolis force balance, producing the geostrophic wind.

The geostrophic wind FIGURE 15.13

zone, belts of high pressure near the Tropics of Cancer and Capricorn, and westerly winds, with some variation in direction, spiraling around polar lows.

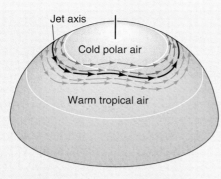

A The flow of air along the front begins to undulate.

Jet axis
Cold polar air
Warm tropical air

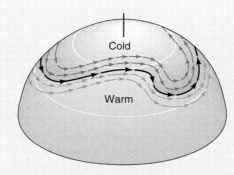

B As the undulation becomes stronger, Rossby waves form. Warm air pushes toward the pole, while cold air is brought to the south.

Cold
Warm

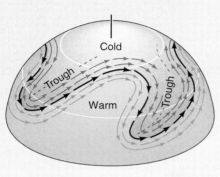

C The waves become stronger, bringing a tongue of cold air to the south as warm air is carried northward.

Cold
Trough
Warm
Trough

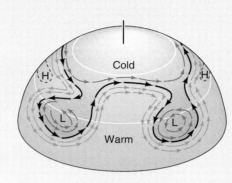

D The tongue is pinched off, leaving a pool of cold air far south of its original location. The waves form cyclones of cold air that may last for up to a week.

H
Cold
H
L
Warm
L

Rossby waves FIGURE 15.14

Rossby waves form at the boundary between cold polar air and warm tropical air.

ROSSBY WAVES, JET STREAMS, AND THE POLAR FRONT

The upper-air westerlies undulate back and forth in giant meanders, which are called **Rossby waves**. These waves arise in the zone where cold polar air meets warm tropical air, called the *polar front*. FIGURE 15.14 describes their formation cycle. Rossby waves are the reason that weather in the midlatitudes is often so variable: They cause pools of warm, moist air and cold, dry air to alternately invade midlatitude land masses.

Jet streams are wind streams that reach great speeds in narrow zones at a high altitude (see *What an Earth Scientist Sees* on page 470). They occur in areas where atmospheric pressure gradients are strong. Along a jet stream, the air moves in pulses along broadly curving tracks. The greatest wind speeds occur in the center of the jet stream, with velocities decreasing farther away from it.

The type of jet stream that is closest to the poles is located along the polar front. It is called the *polar-front jet stream* (or simply the *polar jet*) and is generally located between 35° and 65° latitude in both hemispheres. It follows the edges of Rossby waves at the boundary between cold polar air and warm subtropical air. The polar jet is typically found at altitudes of 10 to 12 km (about 30,000 to 40,000 ft), and wind speeds within it range from 75 to as much as 125 m/s (about 170 to 280 mi/hr).

> **Rossby waves** Horizontal undulations in the flow path of the upper-air westerlies; also known as *upper-air waves.*

> **jet streams** High-speed air flows in narrow bands within the upper-air westerlies and along certain other global latitude zones at high levels.

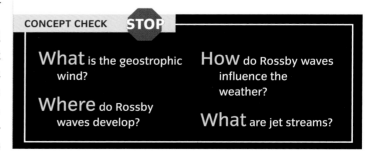

CONCEPT CHECK STOP

What is the geostrophic wind?

Where do Rossby waves develop?

How do Rossby waves influence the weather?

What are jet streams?

Jet Stream Clouds

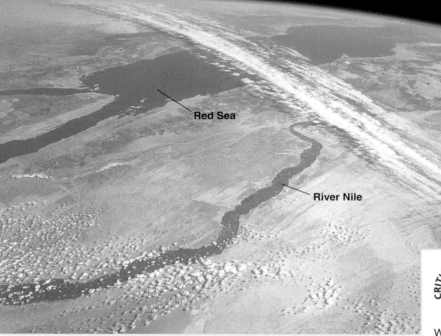

In this photo taken from space, the astronauts aimed their camera to take in the Nile River Valley and the Red Sea. At the left you can see the tip of the Sinai Peninsula. They also captured a beautiful band of cirrus clouds at about 25° north latitude. An Earth scientist would identify this as a band of jet stream clouds that occur on the side of the jet toward the Equator. The jet stream is moving from west to east (from the viewer's right to left) at an altitude of about 12 km (40,000 ft).

Red Sea

River Nile

CRITICAL THINKING

Here's an interesting question:
• To the left of the jet stream clouds there are many small, fluffy clouds. What kind of clouds are they? Are they higher or lower than the jet stream clouds?

Weather Systems

LEARNING OBJECTIVES

Define air mass and **explain** how air masses are classified.

Describe cold, warm, and occluded fronts.

Discuss thunderstorms, wave cyclones, tornadoes, and hurricanes.

W eather systems are often associated with the motion of **air masses**—large bodies of air with fairly uniform temperature and moisture character-istics. An air mass can be several thousand kilometers or miles across and can extend upward to the top of the troposphere. Air masses can be searing hot, icy cold, or any temperature in between, and they can have wildly varying moisture content. They pick up their temperature and moisture characteristics in *source regions*, areas where the air moves slowly or is stationary.

> **air mass** An extensive body of air in which temperature and moisture characteristics are fairly uniform over a large area.

TROPICAL CYCLONES

Tropical cyclones form over warm tropical oceans. They can be very intense storms that devastate islands and coasts. In the Atlantic, they are known as *hurricanes*. In the Pacific, they are *typhoons*, and they are known simply as *cyclones* in the Indian Ocean.

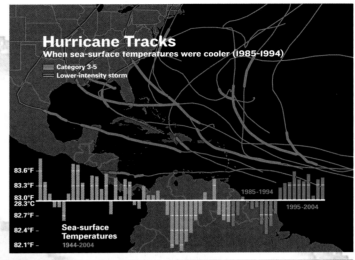

Hurricane Tracks
When sea-surface temperatures were cooler (1985-1994)
- Category 3-5
- Lower-intensity storm

Sea-surface Temperatures
1944-2004

83.6°F
83.3°F
83.0°F
28.3°C
82.7°F
82.4°F
82.1°F

1985-1994
1995-2004

▶ **Hurricane tracks**
Along the southeastern coast of the U.S., many tropical cyclones approach from the east then turn and head back out to sea. These figures show hurricane tracks for 1985–1994 (blue, *upper map*) and 1995–2004 (red, *lower map*). The thicker the line of the track, the more intense the hurricane.

The bar chart in the upper map shows sea-surface temperatures for both periods. Comparing the upper and lower maps, you can see that the number and intensity of storms increases when sea surface temperatures increase.

Now that they're warmer (1995-2004)

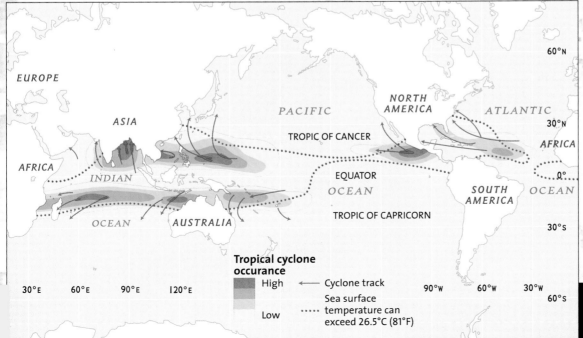

EUROPE

ASIA

PACIFIC

NORTH AMERICA

ATLANTIC

TROPIC OF CANCER

60°N

30°N

AFRICA

AFRICA

INDIAN

EQUATOR
OCEAN

SOUTH AMERICA

OCEAN

0°

OCEAN

AUSTRALIA

TROPIC OF CAPRICORN

30°S

Tropical cyclone occurance
High
Low

←— Cyclone track
····· Sea surface temperature can exceed 26.5°C (81°F)

30°E   60°E   90°E   120°E

90°W   60°W   30°W

60°S

◀ **Oceans and Cyclones**
Tropical cyclones are most likely to occur in areas of greatest heating. Dotted lines show where the sea surface temperature can be greater than 26.5°C (81°F). Cyclones last until they move over cooler waters or hit land. When a cyclone encounters warmer waters, as Hurricane Katrina did in 2005 in the Gulf of Mexico, it picks up energy and intensifies.

481

Amazing Places: New Orleans Before and After Katrina

NATIONAL GEOGRAPHIC

In 2005, Hurricane Katrina laid waste to the city of New Orleans and much of the Louisiana and Mississippi Gulf coasts. Originating southeast of the Bahamas, the hurricane first crossed the South Florida Peninsula as a category 1 storm and then moved into the Gulf of Mexico, where it intensified to a category 5 storm. After the storm weakened somewhat as it approached the Gulf Coast, its eye came ashore at Grand Isle, Louisiana, with sustained winds of 56 m/s (125 mi/hr) early on August 29.

A New Orleans, shown here before Katrina struck, is particularly vulnerable to hurricane flooding because it was built largely on the floodplain of the Mississippi River. Most of its land area has slowly sunk below sea level as underlying river sediments have compacted over time. Levees protect the city from Mississippi River floods, as well as from ocean waters along the saline Lakes Borgne (on the east) and Pontchartrain (on the north).

B Katrina wrought a degree of devastation that was unparalleled in the history of disasters in the United States. Many neighborhoods in New Orleans lay in ruins. This photo was taken about two weeks after the storm. The red object in the foreground is a barge. The devastation is nearly complete. Total losses were estimated at more than $125 billion, and the official death toll exceeded 1800. Adding insult to injury, much of New Orleans was reflooded three weeks later by Hurricane Rita, a category 3 storm that made landfall on September 24 at the Louisiana–Texas border.

SUMMARY

1 Atmospheric Pressure

1. The term **atmospheric pressure** describes the weight of air pressing on a unit of surface area. Atmospheric pressure is measured using a barometer.

2. Atmospheric pressure decreases rapidly as altitude increases.

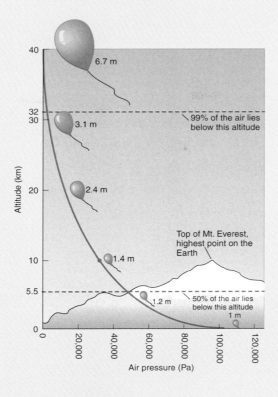

2 Why Air Moves

1. Air motion is produced by **pressure gradients**.

2. Pressure gradients form when air is heated unevenly, creating **convection loops**.

3. Sea and land breezes are examples of convection loops formed from unequal heating and cooling of land and water surfaces. Mountain and valley winds, the Santa Ana, and Chinooks are other examples of local winds.

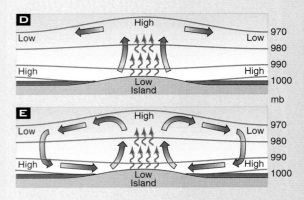

3 Cyclones and Anticyclones

1. Earth's rotation creates the **Coriolis effect**.

2. The *Coriolis force* deflects wind motion, making air spiral around **cyclones** (centers of low pressure and convergence) and **anticyclones** (centers of high pressure and divergence).

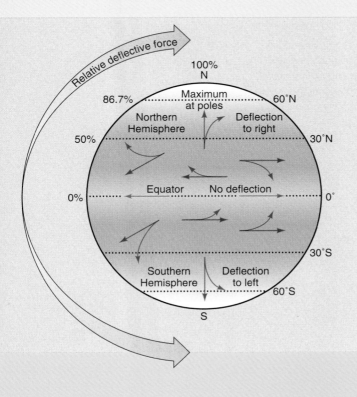

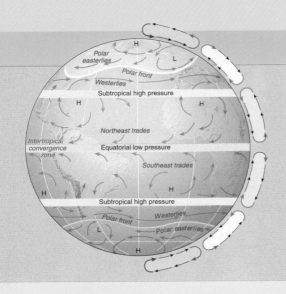

4 Global Wind Patterns

1. Equatorial and tropical regions are heated more intensely than the higher latitudes, setting up convection loops.

2. These loops drive the **trade winds** and the convergence and lifting of air at the **intertropical convergence zone (ITCZ)**.

5 Winds Aloft

1. Winds in the atmosphere are dominated by a global pressure-gradient force between the tropics and the pole in each hemisphere.

2. The global pressure-gradient force and the Coriolis force generate strong westerly **geostrophic winds** in the upper air.

3. **Rossby waves** develop in the upper-air westerlies, bringing cold polar air toward the Equator and warmer air toward the poles. The polar-front and subtropical **jet streams** are concentrated westerly wind streams with high wind speeds.

6 Weather Systems

1. **Air masses** are distinguished by their latitudinal location and their source regions. **Fronts** are the boundaries between air masses. They include cold and warm fronts, where cold or warm air masses are advancing. In the occluded front, a cold front overtakes a warm front, pushing a pool of warm, moist air above the surface.

2. Thunderstorms can form if air is unstable, creating hail and lightning. *Tornadoes* are very small, intense cyclones that occur as a part of thunderstorm activity.

3. **Wave cyclones** are produced when anticyclones meet at the polar front. They form the dominant weather system in the middle and high latitudes.

4. **Tropical cyclones**, also known as *hurricanes* or *typhoons*, develop over very warm tropical oceans and can intensify to become vast inward-spiraling systems of very high winds with very low central pressures.

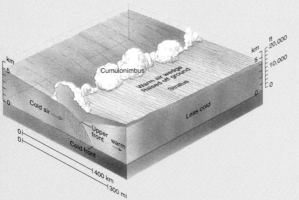

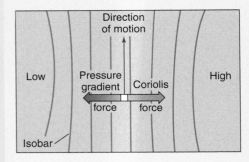

KEY TERMS

CRITICAL AND CREATIVE THINKING QUESTIONS

1. Earth and Venus are almost identical in size and mass, but the atmospheric pressure on the surface of Venus is 90 times greater than air pressure at sea level on Earth. Why?

2. Consult the weather map from your daily newspaper, or download one from a weather station on TV. (Do not look at the weather forecast.) Predict the weather for the week ahead in the place you live or go to school.

3. If air and ocean surface temperatures increase as a result of global climate change, scientists predict that the severity and frequency of hurricanes in the North Atlantic will increase. What reasoning leads them to make such a prediction?

4. Some of the strongest and most sustained winds on record have been recorded at research stations located on the shore at the foot of the Antarctic Ice Cap. What special conditions explain this phenomenon?

5. Imagine you are a commercial airline pilot and you are scheduled to fly nonstop from Vancouver, British Columbia, to Sydney, Australia. What upper-atmosphere wind conditions would you expect to encounter as you travel? What jet streams will you encounter? Will the jet streams slow or speed your aircraft on its way?

What is happening in this picture ?

The photograph shows a line of cumulus clouds advancing from left to right. The clouds were formed when warm, moist air was pushed aloft.

- Why is the warm air forced upward?

- The clouds mark an advancing front. What type of front might this be?

1. The boiling point of water lowers as one goes higher in elevation because _____.
   a. water is less dense at higher elevations
   b. air is denser at higher elevations
   c. air pressure is less at higher elevations
   d. upward water pressure is much greater

2. In the diagram shown, the air will move from the _____ pressure center at _____ to the _____ pressure center at _____.
   a. high/Columbus/low/Wichita
   b. low/Columbus/high/Wichita
   c. high/Wichita/low/Columbus
   d. low/Wichita/high/Columbus

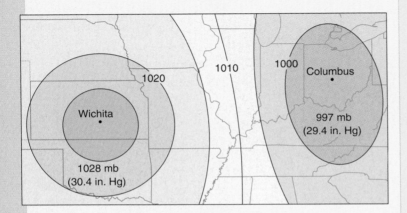

3. A land breeze generally occurs _____.
   a. at night, when the land cools below the surface temperature of the sea
   b. when strong winds blow in from the sea over the land
   c. only during certain restricted seasons
   d. during the day, when the land heats above the surface temperature of the sea

4. A parcel of air at the surface is subjected to three forces and the balance among the pressure gradient, Coriolis, and _____ forces determines the direction of motion of the parcel of air.
   a. gravitational   c. frictional
   b. centrifugal   d. divergent

5. The Coriolis effect is _____.
   a. a result of Earth's rotation from east to west and causes objects to curve to the right in the Northern Hemisphere
   b. a result of Earth's rotation from west to east and causes objects to curve to the right in the Northern Hemisphere
   c. a result of Earth's rotation from west to east and causes objects to curve to the left in the Northern Hemisphere
   d. a result of Earth's rotation from east to west and causes objects to curve to the left in the Northern Hemisphere

6. The diagram shows the motion of air in cyclones and anticyclones. Identify which figures represent: (a) a Northern Hemisphere anticyclone, (b) a Southern Hemisphere anticyclone, (c) a Northern Hemisphere cyclone, and (d) a Southern Hemisphere cyclone.

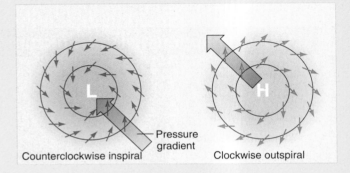

Counterclockwise inspiral   Pressure gradient   Clockwise outspiral

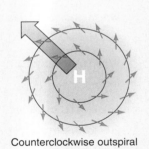

Clockwise inspiral   Counterclockwise outspiral

7. Cloudy and rainy weather is often associated with the inward and upward convergence of air within _____.
   a. anticyclones   c. cold fronts
   b. warm fronts   d. cyclones

8. Label the following features on this illustration of global atmospheric circulation:

intertropical convergence zone
polar front
northeast trade winds
southeast trade winds
subtropical high pressure
polar easterlies
westerlies

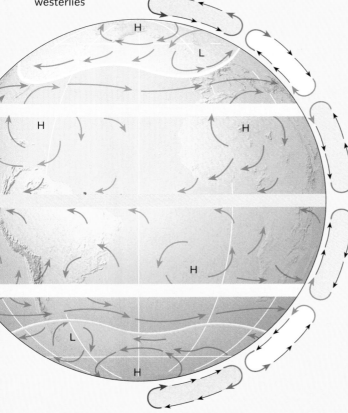

9. At upper levels in the atmosphere, as a parcel of air moves in response to a pressure gradient, it is turned progressively sideways until the gradient and Coriolis forces balance, thus producing the _____.
   a. geostrophic wind
   c. tropospheric wind
   b. upper-air westerlies
   d. equatorial easterlies

10. Jet streams are _____.
   a. narrow zones at a high altitude in which wind streams reach speeds greater than the speed of sound
   b. narrow zones at a high altitude in which wind streams sometimes reach speeds of over 150 miles per hour
   c. rivers of wind that exist only along the Equator and travel at fairly high velocities
   d. rivers of wind that circulate around the poles

11. What type of front is shown here?
   a. cold
   c. occluded
   b. warm
   d. stationary

12. A(n) _____ is a center of high pressure and is generally responsible for fair weather.
   a. anticyclone
   c. trade wind
   b. cyclone
   d. midlatitude storm front

13. A _____ is a small but intense cyclonic vortex with very high wind speeds.
   a. hurricane
   c. tornado
   b. typhoon
   d. cyclone

14. Hurricanes and typhoons generally develop within the _____ latitudinal zones.
   a. 15° to 30° north and south
   b. 10° to 20° north and south
   c. 8° to 15° north and south
   d. 30° to 45° north and south

15. A _____ is a sudden rise of water level caused by a hurricane.
   a. storm surge
   c. flood
   b. tsunami
   d. tidal flood

Global Climates Past and Present

In the late summer of 1991, two German trekkers made a gruesome discovery high in the Tyrolean Alps. The mummified body of a prehistoric man was seen protruding from slowly melting ice near the margin of the Similaun Glacier at 3200 m (10,500 ft) altitude.

Scientific studies have determined that Ötzi the Iceman (see inset), as he has been dubbed, was about 45 years of age when he died. Radiocarbon dating of his skin and bones reveal that he died over 5000 years ago, making him a member of the Late Neolithic and Bronze age population of south-central Europe.

Ötzi's discovery provided unprecedented riches to help archaeologists reconstruct what everyday Neolithic life was like. Along with his corpse were a fur robe, a woven grass cape, leather shoes, a flint dagger, a copper ax, a wooden bow, and 14 arrows.

The discovery was also a boost for climate scientists, confirming the picture of the region they had pieced together from other evidence. At the time that Ötzi lived, the climate was getting cooler and glaciers, such as the now-retreating Bertrab Glacier in South Georgia pictured here, were growing larger. When he died near the margin of the Similaun glacier, his remains became entombed in an Alpine deep-freeze that kept him perfectly preserved for more than 5 millennia. The following years saw a succession of cool periods, the last of which is known as the Little Ice Age. However, none of the intervening mild periods was warm enough to release Ötzi, until the recent warming exposed him to view.

been downward. Viewed in a longer-term context, the current global warming episode will do nothing to change that. Third, in the last 800,000 years the planet has alternated between ice ages, or **glaciations**, and warm *interglacial periods*, each cycle lasting about 100,000 years. In all, more than 20 ice ages have occurred over the past 2 million years. We are currently near the peak of a warm interglacial cycle. In 50,000 years or so, Earth will probably experience another ice age—and the current global warming trend is unlikely to delay or hasten it. At most, it may create a sort of "superinterglacial spike."

> **glaciation** The covering of large land areas by glaciers or ice sheets.

HOW WE KNOW IT:
THE TEMPERATURE RECORD

How do scientists know about Earth's climate and surface temperature from 20,000 years ago, or even 1 million or 100 million years ago? Measured records of air temperatures date back only to the middle of the 19th century. If we want to know about temperatures at earlier times, we need to use indirect methods, such as tree-ring, coral, and ice-core analysis (**FIGURE 16.2**).

Deep-sea sediments provide some of the best indirect evidence we have of past climatic changes (**FIGURE 16.3**). The temperature graph in Figure 16.1 was derived from the isotopes of oxygen in deep-sea fossils. Another important piece of evidence is the extent of glacial deposits such as terminal moraines, described in Chapter 6. Layered sediments also provide information about climate. For example, paleontologists can use plant and animal fossils to infer past climates. Fossilized pollen spores in old bogs and lake-bottom sediments have been particularly useful for reconstructing past climatic changes on a fine scale. Sedimentologists and stratigraphers study ancient soil horizons, called *paleosols*, which represent former land surfaces and provide information about climate and weather at the time they formed. The minerals present in paleosols allow geochemists to determine the chemical composition of the ambient air and water.

As you can see, the evidence of past temperature fluctuations and glaciations comes from multiple sources. No single source is definitive.

Deep-sea microfossils FIGURE 16.3

A Seafloor sediments obtained by drilling contain fossils of tiny sea creatures, foraminifera, that once lived in surface waters. These microfossils speak volumes about the chemistry and temperature of the ocean. Foraminifera have calcium carbonate shells that contain oxygen.

B The ratio of oxygen-16 (the lighter isotope of oxygen) to oxygen-18 is a measure of the temperature of the seawater in which the microscopic creatures lived.

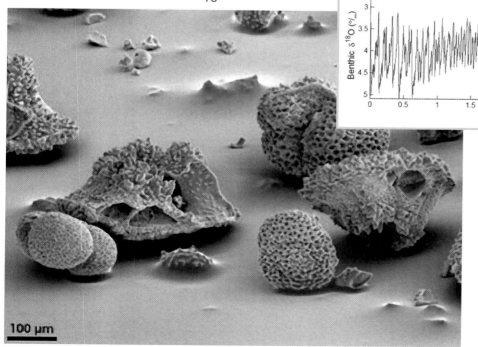

The Gubbio Sediments

Rock near the Italian town of Gubbio reveals secrets about changes in Earth's orbit over millions of years. The surrounding hills expose thousands of thin layers of limestone, and a sedimentary rock that is rich in clay and organic carbon, called *marl* (A). An Earth scientist looking at the light and dark stripes would realize that for tens of millions of years the seafloor chemistry cycled between an oxygen-rich and an oxygen-poor environment. These sediments accumulated from a slow rain of dead marine organisms. When dissolved oxygen was present, scavenger organisms picked the skeletal material clean, leaving behind calcium carbonate to form limestone. Without oxygen, organic carbon was preserved, leaving a dark marl.

Earth scientists have found that these variations match the signature of Milankovitch cycles (B). These alter the pattern of solar heating by just a few percentage points, comparable to the recent heating due to greenhouse gases. The rock record shows that when pushed, Earth's climate can jump.

A

B Three kinds of orbital change affect Earth's climate. When these three factors are added together, they greatly affect how much sunlight reaches Earth, and where, at any given time.

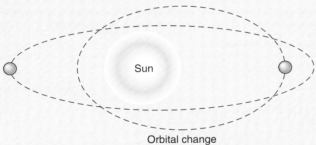

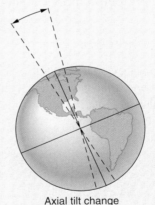

Orbital change
Earth's orbit becomes more and less elongated over a period of 100,000 years.

Axial tilt change
Earth's axis changes its tilt over a period of 41,000 years.

Wobble
Earth's axis wobbles in a circle once every 26,000 years.

Here's an interesting question:
• It was a scientist studying the cycling of the sedimentary layers who first noticed a thin layer that differed from all the rest. Research showed it to be a layer of debris from a meteorite impact. This was the first clue that led to the hypothesis of dinosaur extinction due to a giant impact. How likely is it that such a thin layer—no more than a few centimeters—would actually be noticed?

CAUSES OF CLIMATE CHANGE

It is clear that climate has changed dramatically in the past and will continue to do so in the future. However, the reasons for climate change are somewhat murkier, and this makes it hard for us to predict *how* the climate will change and at what rate.

Several mechanisms cause natural climate changes. Some of these are geographic changes. For example, during the Cenozoic Era, tectonic activity created the Isthmus of Panama, which joins North and South America. This land bridge severed the connection between the Atlantic and Pacific Oceans, altered oceanic and atmospheric circulation, and thus had a major impact on global climate. Other factors include changes in solar activity and volcanic eruptions. As solar output increases, global temperatures rise. Volcanic gases released into the atmosphere can condense into tiny liquid droplets, known as *aerosols*, that can have a cooling effect; aerosols from other sources, such as forest fires or power plants that release carbon particles, can sometimes have a warming effect.

Astronomical factors are also believed to affect Earth's climate. Small changes in the *eccentricity* (departure from circularity) of Earth's orbit, the *tilt of* the planet's axis of rotation, and the *precession* (wobbling) of the axis affect how much solar radiation reaches Earth's surface and at what times of the year. These variations, called **Milankovitch cycles**, correspond reasonably well to the periods of past glaciations (see *What an Earth Scientist Sees*). The combined effect of tilt and precession is roughly correlated with 20,000- to 40,000-year interglacial cycles, and variations in eccentricity may contribute to cycles that last 100,000 years. Earth's orbit is discussed in more detail in Chapter 17.

> **Milankovitch cycles** Climate cycles that occur over tens to hundreds of thousands of years because of changes in Earth's orbit and tilt.

All these influences on climate are tempered by the *greenhouse effect*, which we discussed in detail in Chapter 14. Water vapor, carbon dioxide, and methane in the atmosphere store heat that radiates from Earth's surface and radiate it back downward. Without this natural greenhouse warming by the atmosphere, the average temperature on the surface would be much cooler.

RAPID CLIMATE CHANGE: THE YOUNGER DRYAS EVENT

At the end of the last glaciations, about 11,000 to 10,000 years ago, the climate in the North Atlantic and adjacent lands experienced a rapid and remarkable change (**FIGURE 16.4**). For 2000 years, the

---

Evidence for the Younger Dryas event FIGURE 16.4

A Under full-glacial conditions, plants that are currently limited to polar and high-altitude regions could move into forests in northwestern Europe. Among these plants is *Dryas octopetala*, shown here. A large amount of *Dryas* pollen was found in organic deposits dating to the cold period, now known as the Younger Dryas event.

B Measurements of oxygen isotopes in sediments taken from a Swiss lake and an ice core from the Greenland Ice Sheet show that both the onset and the end of the Younger Dryas event were rapid. You can view the curves in the graph as marking changes in temperature. At the end of the event, the climate over Greenland warmed by about 7°C in only 40 years—a rate that exceeds that predicted by climate models for the coming century.

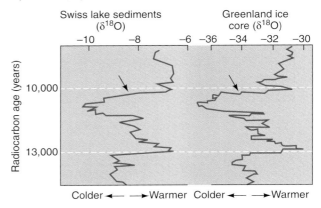

climate had been warming, causing ice sheets in North America and Europe to retreat and allowing plants and animals to reoccupy deglaciated land. Then, very abruptly, the climate cooled. Water temperatures in North America fell, and the ice sheets stopped retreating and began advancing again. This cold episode is known as the **Younger Dryas** event.

■ **Younger Dryas** A cold period that occurred between about 11,000 and 10,000 years ago, during the generally mild epoch.

What caused such a dramatic and rapid change in climate? The effects of Younger Dryas cooling are most pronounced around the North Atlantic, so it makes sense to look to that region for clues. It turns out that the solution lies in the interactions between the melting ice sheets and ocean circulation patterns, as described in FIGURE 16.5.

Process Diagram

VIEW THIS IN ACTION
in your WileyPLUS course

Causes of the Younger Dryas event FIGURE 16.5

Ocean circulation in the North Atlantic plays an important role in controlling climate, as we saw in Chapter 12, where we met the thermohaline circulation system, also known as the *ocean conveyor belt*.

1 Meltwater lakes As the ice sheet over eastern North America retreated, vast meltwater lakes holding icy water were created. At the same time, the ocean conveyor belt was at work in the North Atlantic. Wind-driven warm surface currents, such as the Gulf Stream, headed toward the poles, cooling and eventually sinking at high latitudes, having transferred heat energy around the globe.

2 Drainage As the ice shrank further, it uncovered a natural drainageway between the meltwater lakes and the North Atlantic. The meltwater flooded rapidly into the ocean, forming a freshwater lid over the denser salty seawater. The cold surface meltwater in turn reduced the salinity of the water and the rate of evaporation from the ocean surface, shutting down the normal pattern of ocean circulation, known as the *thermohaline conveyor system*.

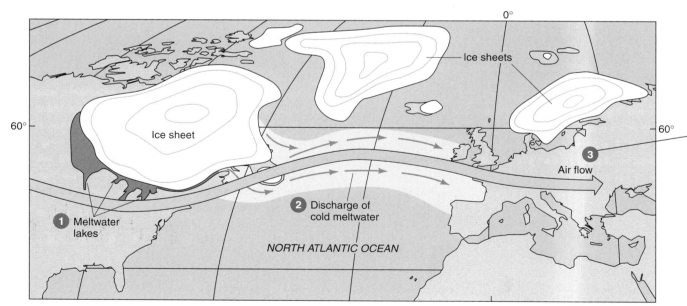

Global warming due to climate change will have repercussions around the world.

◀ Asia

Crop production in northern and midlatitude Asia will improve, thanks to the warmer weather. However, South and Southeast Asia's many developing countries will suffer from intolerably high temperatures that will reduce food production and from a drop in rainfall and water supply. In arid and semiarid Asia, higher temperatures and more evaporation will reduce rice yields dramatically.

◀ Small Island States

The tens of thousands of small islands scattered across the world's oceans are particularly vulnerable to climate change. Many rise only 1 or 2 m above sea level. Male, the capital of the Maldives, could find itself underwater if sea levels rise. The Maldives, an island nation in the Indian Ocean, are built on coral reefs that grow over time but cannot possibly grow fast enough to keep up with a rapid change in sea level.

◀

Australia and New Zealand

Crops in arid Australia are already growing near their maximum heat tolerance, so warmer weather will affect food production in these regions. New Zealand, which is cooler and wetter, could initially benefit from warmer temperatures.

◀ Africa

Over the next century, East Africa could receive more rain, but southern Africa is forecast to become much drier. Floods and storms will increase in frequency, and food and water shortages will be a growing problem.

Amazing Places: Barrow, Alaska

Global Locator

NATIONAL GEOGRAPHIC

A

B

The people of Barrow, Alaska, live in one of the most extreme environments on our planet (**A**). Situated 539 km north of the Arctic Circle, this is the northernmost settlement on the North American mainland. In winter, the Arctic Ocean freezes right up to the coast, and when the sun sets in mid-November, it does not rise again until January.

The Inupiat inhabitants are acutely vulnerable to climate change. They travel out on the sea ice in spring and summer to hunt whales, seals, and walrus for sustenance (**B**). However, as weather conditions warm, the amount of sea ice is decreasing, and hunting is becoming increasingly precarious. The way that families store food is also affected, as higher temperatures make it harder to store food in natural freezers below the ground.

The poles are warming much faster than the rest of the planet. Average temperatures in Barrow have increased by around 2.5°C in the past 30 years. Earth scientists long ago predicted that the most visible impacts from a globally warmer climate would strike high latitudes first. They pre-dicted that air and sea temperatures would rise, that the snow would melt earlier and the ice would freeze later, and that there would be an increase in storm intensity. Now all of those impacts have been documented in Alaska, which stands as an early warning system for the rest of the planet.

1 Global Climate Change in the Past

1. **Climate** change is nothing new. Over the past 2 million years, Earth has experienced repeated **glaciations** and warm interglacial periods. We are now in an interglacial period. The last glaciation, or ice age, peaked about 18,000 years ago. The current trend of global warming may create a more pronounced interglacial period but is unlikely significantly to delay the next ice age, which can be expected in 50,000 years or so.

2. There are direct records of air temperature dating to the middle of the 19th century. To reconstruct past climates beyond that, Earth scientists analyze fossils, sediments, tree rings, corals, and ice cores. The composition of the shells of microscopic fossils, found in ocean sediments, reflects changes in the proportions of warm water and cold water. Fossilized plants also reveal changes in world temperatures. Ice cores preserve trapped bubbles of "fossil air" that help

scientists determine the concentration of greenhouse gases in the atmosphere in the past.

3. Climate change can come about naturally for many reasons, making it difficult to predict exactly how the climate will change. Tectonic activity can shift land masses, leading to changes in ocean and atmosphere circulation. Volcanic activity can release aerosols into the air, which affect the amount of heat absorbed and emitted in the atmosphere. Changes in solar output and astronomical cycles—called **Milankovitch cycles**—also influence how much solar energy reaches Earth. Greenhouse gases have always affected the climate.

4. Climate can also change rapidly. During the **Younger-Dryas** event, which occurred between 10,000 and 11,000 years ago, the climate in the North Atlantic abruptly cooled. This happened because

melting glaciars shutdown an important ocean circulation system, the thermohaline circulation.

2 Global Climates Today

1. Climate depends on latitude, location, and moisture. Latitude determines the annual pattern of insolation; near the Equator, temperatures are warmer and toward the poles, temperatures are colder. Location—continental or maritime—can enhance or moderate the annual insolation cycle because ocean-surface temperatures vary less with the seasons than land-surface temperatures. Air temperature also has an important effect on precipitation because warm air can hold more moisture than cold air.

2. Global precipitation patterns are determined largely by **air masses** and their movements. The latitude at which the air

mass originates determines its temperature. Its original coastal or continental location determines its moisture content. Cyclonic precipitation is created at the *frontal zones* where air masses meet.

3. The globe can be split into three broad climate bands, arranged by latitude. Low-latitude climates (Group I) are typically warm and can range from extremely wet to severely dry. **Low-latitude rainforests** are found in this climate group. Midlatitude climates (Group II) are more variable because conflicting air masses tend to meet and interact in these regions. High-latitude climates (Group III) become colder and dryer approaching the poles.

3 Present-Day Climate Change

1. The possibility of **anthropogenic** climate change due to the burning of fossil fuels has arisen only in the past century. Ice core analysis suggests that carbon dioxide levels are rising much more rapidly than at any time in the past 100,000 years. Such a rise can be accounted for by the 8 billion tons of carbon that humans pump into the atmosphere each year. Many scientists believe that global warming will continue through the 21st century and that its extent will depend on human actions to limit the production of greenhouse gases.

2. So far, the effects of the warming are most pronounced in the polar regions, where glaciers are retreating and ice caps are melting.

3. If fossil-fuel consumption continues to grow at an increasing rate, atmospheric carbon dioxide levels are projected to double by 2100 and average global temperatures are predicted to rise by between 1.5° and 4.5°C. Worldwide effects in the future will vary according to location but may include rising sea levels, increases in the severity of weather systems, and changes in animal habitats that will cause some species to die and force others to migrate to new territory.

KEY TERMS

- **climate** p. 491
- **glaciation** p. 493
- **Milankovitch cycles** p. 495
- **Younger Dryas** p. 496
- **low-latitude rainforest** p. 500
- **anthropogenic** p. 505

CRITICAL AND CREATIVE THINKING QUESTIONS

1. If you were to set out to make a model that might be used to predict global climates 50 years into the future, what kind of climate data would you put into your model? Once you have developed your model, how might you test it?

2. Suppose Earth's axis was perpendicular to the plane of Earth's orbit around the Sun. How would the climate at the place you live differ from the climate today? How would it change if Earth's orbit became much more elliptical?

3. Is there any evidence in the landforms around the area in which you live or go to school to indicate that the area was glaciated during the most recent ice age? How thick was the ice at maximum glaciation?

4. There is evidence in the ice cores that there have been very rapid increases and decreases in temperatures in the past. Suppose that climate modelers predict that 50 years from now there will be a time of rapid temperature rise of as much as 5°C over 10 years. Choose one country from each of the three climate groups discussed in the chapter and suggest what changes they should expect.

5. Find out whether your city, state, province, or country has set goals for the reduction of carbon dioxide emissions in order to limit their potential contribution to global warming. What steps have been taken to meet these goals?

What is happening in this picture ?

The photo shows bubbles of air locked in an ice core extracted from the rapidly retreating Quelccaya ice cap in Peru. As ice in a glacier recrystallizes, tiny bubbles of air become permanently trapped in the ice. When the ice is melted under controlled conditions in the laboratory, this fossil air can be analyzed.

- What measurements would you ask scientists to make?

- What information could they gain from them?

1. Study of Earth's climate record indicates that temperatures have _____ over the past two million years.
   a. fluctuated greatly
   b. remained constant
   c. been slowly decreasing
   d. been slowly increasing

2. From the standpoint of global climate, Earth is now _____.
   a. at the beginning of an interglacial period
   b. at the end of a glaciation
   c. at the peak of an interglacial period
   d. 20,000 years into a superinterglacial period

3. Earth scientists use a number of techniques to study paleoclimate, including data collected from _____.
   a. fossil pollens from peat bogs and lakes
   b. stable oxygen isotope ratios
   c. ice cores
   d. All of the above are techniques used to study paleoclimate.

4. These microfossils help scientists study climate change because they record changes in the concentration of _____ in the ocean, which can be linked to changes in temperature.
   a. calcium carbonate   c. carbon dioxide
   b. oxygen   d. hydrogen

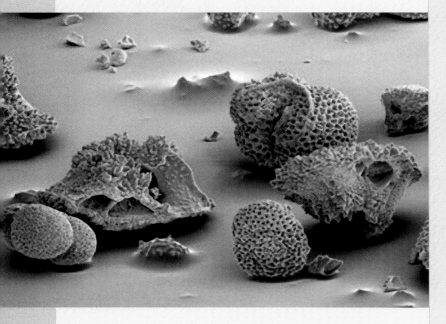

5. Earth science studies have revealed a number of natural agents that can act together to cause climate change. Which of the following are considered possible causes of climate change?
   a. changes in the eccentricity of Earth's orbit
   b. plate tectonic activity
   c. interruptions of ocean circulation
   d. the tilt and precession of Earth's rotational axis
   e. All of the above are considered possible natural causes of climate change.

6. The layers in this rock provide evidence of _____.
   a. the Younger-Dryas event
   b. superinterglacial spikes
   c. Milankovitch cycles
   d. tipping points

7. Approximately when did the Younger Dryas event occur?
   a. 5000 years ago   c. 50,000 years ago
   b. 10,000 years ago   d. 100,000 years ago

8. Which of these is not a key factor in controlling climate?
   a. coastal–continental location
   b. plate tectonic activity
   c. latitude
   d. longitude

9. Name the three climate groups and sketch their rough boundaries on the diagram.

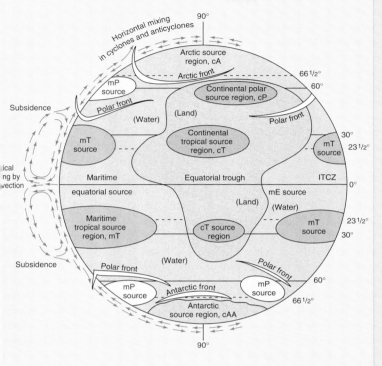

10. In which climate group would you expect to find these trees?
a. Group I
b. Group II
c. Group III

11. In which climate group would you find the Mediterranean climate and the tundra (respectively)?
a. Group I and Group II
b. Group I and Group III
c. Group II and Group III
d. Group III and Group I

12. Anthropogenic carbon dioxide in the atmosphere results from the continued burning of fossil fuels. How far back does the recent record of atmospheric carbon dioxide measurements extend?
a. 100 years
b. 75 years
c. 50 years
d. 10 years

13. Atmospheric carbon dioxide is projected to double by _____, if not sooner, leading to a predicted global temperature rise of between _____° and _____°C.
a. 2050/3.5/7.5
b. 2070/2.5/5.0
c. 2100/1.5/4.5
d. 2200/2.5/4.5

14. What effects would continued global warming have over the next century?
a. retreat of glaciers
b. calving of large icebergs from ice shelves
c. shrinking of some animal habitats
d. expansion of some animal habitats
e. All of the above are possible effects of global warming.

15. Food production in South and Southeast Asia is predicted to _____ as rainfall _____, affecting rice yields. In North and mid-latitude Asia, food production will _____.
a. increase/increases/increase
b. increase/decreases/decrease
c. decrease/increases/decrease
d. decrease/decreases/increase

Earth's Place in Space

In 2006, our solar system received a makeover: Pluto was demoted from its position as the ninth planet and reclassified as a dwarf planet.

Pluto's story begins with a case of mistaken identity. In the 1840s, Urbain Le Verrier predicted the position of the then-undiscovered Neptune, after analyzing unexpected deviations in the orbit of Uranus that could be explained by the gravitational tug of a nearby massive object. Soon after, Neptune was indeed discovered. When astronomers noticed that its orbit seemed to be similarly disrupted, they began to hunt for another new planet, and on February 18, 1930, Pluto was discovered by Clyde Tombaugh. As it turned out, Pluto's tiny mass was too small to significantly disrupt Neptune, and it was later found that, in fact, no missing planet is needed to explain Neptune's orbit. However, Pluto, pictured in this Hubble Space Telescope image surrounded by three moons, took its place as the ninth planet.

Nevertheless, Pluto never quite fit in. Lying farthest from the Sun, it is little larger than the Moon and is dwarfed by its neighbors, Jupiter, Saturn, Uranus, and Neptune. Its fate was sealed with the discovery in 2005 of Eris, an object slightly bigger than Pluto that also orbits the Sun. Rather than officially inducting Eris as the tenth planet—and risk opening the planetary club to many other such small bodies—the International Astronomical Union created a new definition of *planet*. This included the criterion that the body must have cleared comparable-sized objects from the neighborhood of its orbit, and both Pluto and Eris failed this condition.

Pluto

Nix

Charon

Hydra

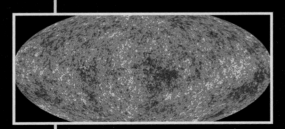

Astronomy and the Scientific Revolution

For much of human history, people believed that everything in the universe revolved around Earth—for good reason. Observing the Sun rising and setting, while Earth seemingly stood still, they reasoned that the universe was **geocentric**, with the Sun revolving around a stationary Earth. Today we are taught that Earth revolves around the Sun. But just how do we know that? The search for evidence that we live in a **heliocentric** universe was a major factor in the rise of modern science.

■ **geocentric** A model of the universe in which a stationary Earth is at the center and all other celestial bodies revolve around it.

■ **heliocentric** A model of the universe in which a stationary Sun is at the center and the planets revolve around it.

IDEAS FROM ANTIQUITY

Ancient Greek civilization, which flourished for 800 years from about 650 B.C. to A.D. 150, spawned many famous philosophers. One of the most influential of these, Aristotle (384–322 B.C.), favored the notion of a geocentric universe (**FIGURE 17.1**). However, a few others, most notably Aristarchus (312–230 B.C.), realized that the daily rising and setting of the Sun and the apparent movement of the star sphere across the sky could be explained just as well if the stars were fixed and Earth rotated on its axis once every day. Using two of the branches of mathematics discovered by the Greeks, geometry and trigonometry, Aristarchus determined that the Sun was huge compared to a relatively small Earth and tiny Moon. The idea of a huge Sun revolving around a small Earth did not make sense to him. In addition, he noticed that a heliocentric universe helped to explain the origin of the seasons, which would arise naturally if the axis of rotation was tilted and Earth revolved around the Sun. But

The celestial spheres FIGURE 17.1

Aristotle pictured the Sun, the Moon, the five visible planets, and the stars as being suspended on concentric, hollow spheres that rotate about an imaginary axis extending outward from the two poles of Earth. Beyond the star sphere lay the realm of the gods.

Aristarchus could not win others over to his heliocentric suggestion, and by the 16th century a geocentric universe had come to be accepted as a divine fact—to be protected by the Catholic Church.

COPERNICUS'S CHALLENGE

The hardest challenge for the concept of a geocentric universe was explaining the motions of the planets. To the naked eye, the five visible planets—Mercury, Venus, Mars, Jupiter, and Saturn—look like stars, except that they seem to wander in relation to the fixed stars. Indeed, the very name *planet* comes from *planetai*, an ancient Greek word meaning "wanderers." The paths followed by these wanderers are odd ones. The planets move a bit farther east each evening, but periodically they slow down and briefly reverse direction before resuming their eastward motion. The temporary reversal of direction is known as *retrograde motion*.

During the 1490s, Nicolaus Copernicus (1473–1543), a student at the University of Bologna, Italy, realized that a heliocentric system could easily explain retrograde motion. The appearance of retrograde motion would result from differences between the time it takes Earth to orbit the Sun and the time it takes any other planet to orbit the Sun, as shown in **FIGURE 17.2**. Moreover, Copernicus suggested that because Mars has a larger retrograde motion than does Jupiter or Saturn, it must be the closest of the three planets to Earth, while Saturn, with the smallest retrograde motion, must be the

most distant. This proposal could be tested by suitable astronomical measurements.

Copernicus also offered two other major hypotheses. First, he suggested that the positions of the planets at any given time in the future could be accurately predicted by assuming that they move in circular orbits around the Sun. Second, he revived the old suggestion of Aristarchus that Earth spins on its axis and the daily

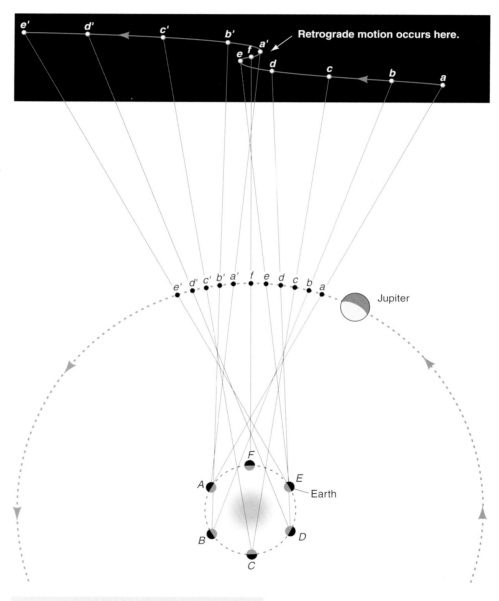

Retrograde motion FIGURE 17.2

As Jupiter moves from point *a* to point *e'* in its orbit, Earth moves counterclockwise from point *A*, completely around the Sun and back to *A*, and then on to point *E*. When one watches Jupiter from Earth, the planet seems to wander in relation to the fixed stars.

rising and setting of the Sun is a result of that rotation. It took until 1851 to demonstrate that Earth really does spin on its axis; this was accomplished using Foucault's pendulum, as described in **FIGURE 17.3**. By extension, Copernicus could explain the seasonality of Earth's climate (**FIGURE 17.4**).

Copernicus's view of the universe directly challenged the Roman Catholic Church, which had built the concept of a geocentric universe into Church doctrine. Looking back, we can see that Copernicus sowed the seeds that finally separated science from religion,

Foucault's pendulum FIGURE 17.3

Léon Foucault provided the first demonstration of the rotation of Earth in 1851 using a huge pendulum, with a 28 kilogram bob and a 67 meter wire, hung from the dome of the Panthéon in Paris. The pendulum was free to swing in any vertical plane. Because Earth rotated under the pendulum, the direction along which the pendulum swung appeared to shift, rotating clockwise 11° per hour and making a full circle in 32 hours.

Process Diagram

The seasons: Earth–Sun relations through the year FIGURE 17.4

VIEW THIS IN ACTION
in your WileyPLUS course

The four seasons occur because Earth's axis is tilted by 23 1/2°.

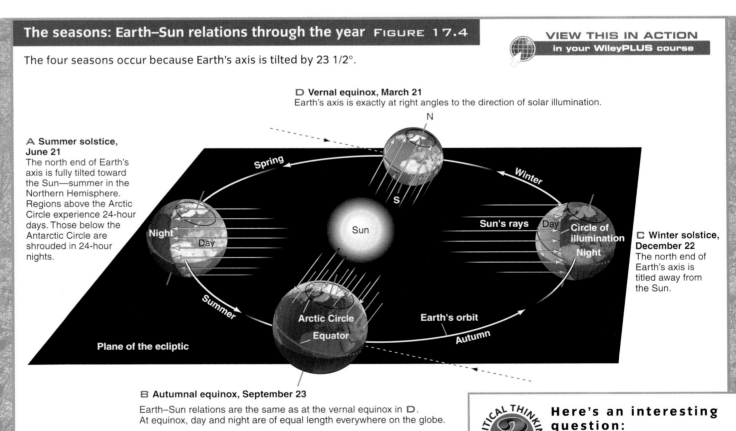

D Vernal equinox, March 21
Earth's axis is exactly at right angles to the direction of solar illumination.

A Summer solstice, June 21
The north end of Earth's axis is fully tilted toward the Sun—summer in the Northern Hemisphere. Regions above the Arctic Circle experience 24-hour days. Those below the Antarctic Circle are shrouded in 24-hour nights.

C Winter solstice, December 22
The north end of Earth's axis is tilted away from the Sun.

B Autumnal equinox, September 23
Earth–Sun relations are the same as at the vernal equinox in **D**.
At equinox, day and night are of equal length everywhere on the globe.

CRITICAL THINKING

Here's an interesting question:
• How would the seasons differ in the place where you live if Earth's axis were perpendicular to the plane of its orbit around the Sun?

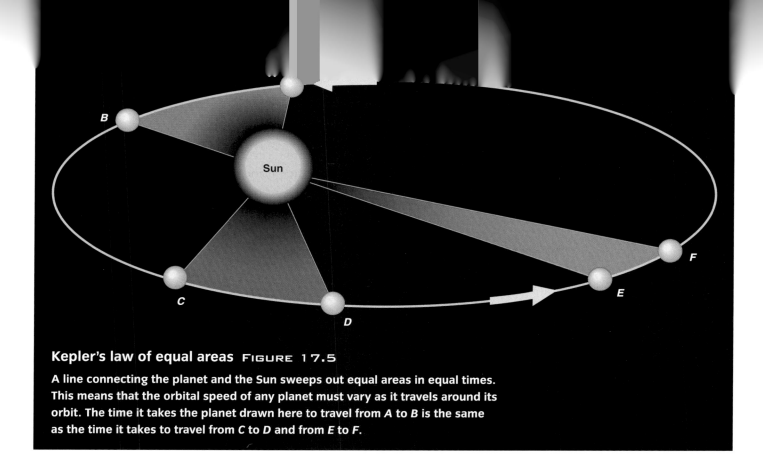

Kepler's law of equal areas FIGURE 17.5

A line connecting the planet and the Sun sweeps out equal areas in equal times. This means that the orbital speed of any planet must vary as it travels around its orbit. The time it takes the planet drawn here to travel from *A* to *B* is the same as the time it takes to travel from *C* to *D* and from *E* to *F*.

spawning the scientific revolution that shaped the society in which we live today.

KEPLER AND THE NEW ASTRONOMY

Although Copernicus was correct in proposing that we live in a heliocentric universe, he was wrong about the shape of the planets' orbits. But it took years of painstaking naked-eye measurements of planetary positions to reveal his mistake. In 1572, Tycho Brahe (1546–1601) built the first modern astronomical observatory on the Danish island of Hven. Tycho was skeptical of the heliocentric hypothesis, and to prove Copernicus wrong he began collecting the most accurate planetary data up to that time.

In 1597, Tycho hired a young German mathematician, Johannes Kepler (1571–1630), to carry out astronomical calculations. Kepler, unlike Tycho, thought Copernicus might be right; he also gave a great deal of thought to a problem Copernicus had not treated: What is the nature of the force that keeps the planets moving around the Sun? Why do they revolve in orbits instead of moving in straight lines out into space?

Because the planets closest to the Sun move faster than those far away, Kepler suggested that a mysterious force must reside in the Sun and have a greater effect on closer objects. Today we know that the force is gravity, but in Kepler's day gravity was an unknown concept. Kepler suggested that the force might be magnetism.

Try as he might, Kepler could not make planetary positions calculated from circular orbits agree with Tycho's measurements. Eventually he tried calculating the position of Mars based on an elliptical orbit, with far better success.

Kepler discovered three laws that describe planetary motion:

1. *The law of ellipses.* The orbit of each planet is an ellipse with the Sun at one focus.
2. *The law of equal areas.* A line drawn from a planet to the Sun sweeps out equal areas in equal amounts of time (FIGURE 17.5). Because of this, the orbital speeds of the planets are not uniform. A planet moves rapidly when close to the Sun and slowly when far away from the Sun.

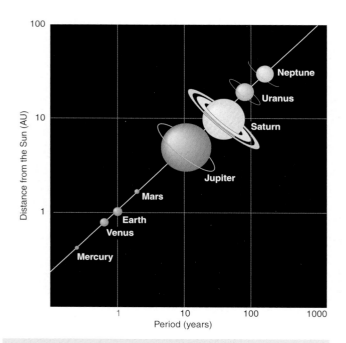

Kepler's law of orbital harmony FIGURE 17.6

Kepler noted that the distance of the planet from the Sun is related to the period of the planet's orbit around the Sun. The unit of distance is the *astronomical unit*, the average distance of Earth from the Sun. You can see that there is a gap between Mars and Jupiter. This is where the asteroid belt lies. The asteroid belt contains small, rocky fragments that did not cluster together to form planets.

3. *The law of orbital harmony.* For any planet, the square of the orbital period in years is proportional to the cube of the planet's average distance from the Sun (FIGURE 17.6). The period is the time a planet takes to make one complete revolution around the Sun. For example, the period of Earth is 365.24219 days.

GALILEO AND NEWTON

Galileo Galilei (1564–1642) was an extraordinary man who made a great many scientific discoveries. In 1609, he constructed a small telescope that magnified objects by 30 times. With this device, he helped put to rest the notion of a geocentric universe. Among other discoveries, Galileo saw four moons orbiting Jupiter, proving that Earth is not the center of all orbital motion. He also found that Venus has phases, just like the Moon, and also changes greatly in apparent size—a fact that can be explained only if Venus and Earth are in orbit around the Sun (FIGURE 17.7).

Galileo also made major contributions to our understanding of moving bodies and gravity. He reasoned that bodies move when they are acted on by forces, and once a body is moving, it will stop or change direction only if another force is applied to it. He also concluded that gravity pulls all falling bodies with the same accel-

Phases of Venus
FIGURE 17.7

When Earth and Venus are on the same side of the Sun, Venus is seen as a crescent. When Earth and Venus are on opposite sides of the Sun, Venus is seen as a full disk, but it is only one-seventh the diameter of the crescent because it is so far away. In a geocentric system, with Venus in orbit around Earth, the apparent size of Venus should change very little.

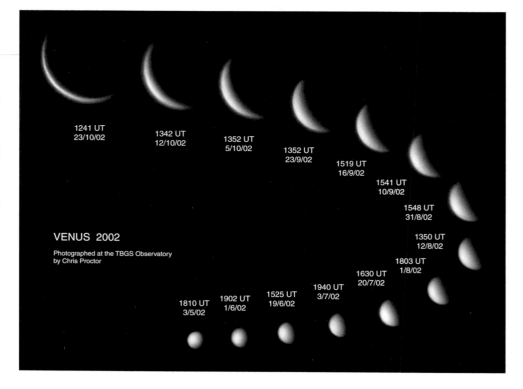

Earth's pull on the Moon FIGURE 17.8

A "Earthrise" seen from the Moon by the astronauts on the Apollo 8 spacecraft in 1968. Newton realized that the same force of gravity that holds us on the ground also extends out to the Moon.

Direction of motion in the absence of a gravitational force

Earth

Gravitational force

Moon

B Gravity is constantly diverting the Moon's direction of motion from a straight line to a closed ellipse.

eration, regardless of their mass. His insight provided one of the key steps to explaining the law of gravity and the motions of planets. The person who pulled all the pieces together, however, was Isaac Newton (1642–1727).

According to legend, Newton started thinking about gravity when he saw an apple fall from a tree. He reasoned that if the force of gravity acts on an apple, that force must also extend "to the orb of the Moon." The Moon revolves around Earth instead of moving through space in a straight line because the force of Earth's gravity exerts a small pull on it (FIGURE 17.8). The force that Kepler had misidentified as magnetism was actually gravity. Newton had discovered that gravity acts between all bodies.

With his discovery, Newton had finally united the heavens, which once were believed to belong to the gods, with Earth, showing that the same forces are at play in both realms.

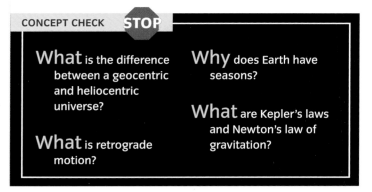

CONCEPT CHECK STOP

What is the difference between a geocentric and heliocentric universe?

What is retrograde motion?

Why does Earth have seasons?

What are Kepler's laws and Newton's law of gravitation?

The Solar System

LEARNING OBJECTIVES

Explain the nebula hypothesis and the origin of the solar system.

Describe the structure of the Sun and **explain** how it produces energy.

Identify the characteristics of the terrestrial and Jovian planets.

Describe meteorites, asteroids, comets, and the role of impacts in the formation of the Moon.

Discuss other solar systems and the possibility of life on other planets.

J ust how did the solar system that we call home form? Any theory for the origin of the solar system must be able to explain its two most striking characteristics. In

Chapter 1, we met the first of these features: The planets can be separated into two distinct groups based on their density and closeness to the Sun (see Figure 1.9). The terrestrial planets—Mercury, Venus, Earth, and Mars—lie closest to the Sun and are small, dense, and rocky. The asteroids, which orbit the Sun in the asteroid belt, as we saw in Figure 17.6, are also rocky, dense bodies, but they are too small to be called planets. The Jovian planets lie farther from the Sun than Mars and are much larger than the terrestrial planets, yet much less dense.

If you were to view the solar system from a spaceship, you would notice a second remarkable feature: Almost everything—planets, moons, Sun—revolves and rotates in the same direction. All the planets revolve around the Sun and all the moons revolve around their respective planets in approximately the same plane. If your vantage point is high above the North Pole, the

Process Diagram

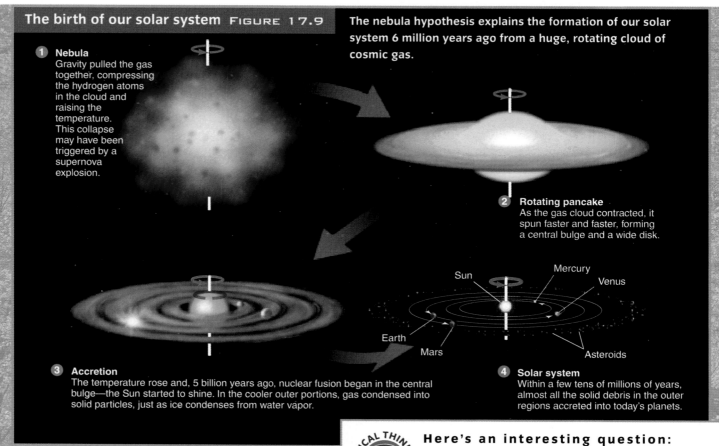

The birth of our solar system FIGURE 17.9

The nebula hypothesis explains the formation of our solar system 6 million years ago from a huge, rotating cloud of cosmic gas.

1 Nebula
Gravity pulled the gas together, compressing the hydrogen atoms in the cloud and raising the temperature. This collapse may have been triggered by a supernova explosion.

2 Rotating pancake
As the gas cloud contracted, it spun faster and faster, forming a central bulge and a wide disk.

3 Accretion
The temperature rose and, 5 billion years ago, nuclear fusion began in the central bulge—the Sun started to shine. In the cooler outer portions, gas condensed into solid particles, just as ice condenses from water vapor.

Sun — Mercury — Venus — Earth — Mars — Asteroids

4 Solar system
Within a few tens of millions of years, almost all the solid debris in the outer regions accreted into today's planets.

CRITICAL THINKING

Here's an interesting question:
• Why don't the following negate the nebular hypothesis: Venus rotates slowly in the opposite direction to the other planets, Earth's axis is tilted to the plane of its orbit, and Uranus's axis is near the plane of its orbit?

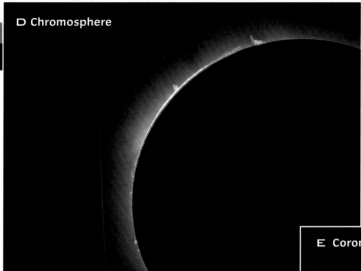

D Chromosphere

The Sun's outermost atmospheric layers—**D**, the *chromosphere* and **E**, the *corona*—can be seen only during a total solar eclipse, when light from the brilliant photosphere is blocked.

E Corona

F Auroras are produced when charged particles from the solar wind collide with particles in Earth's atmosphere.

THE PLANETS

As we saw earlier, the planets in our solar system can be divided into two groups: the innermost terrestrial planets, and the outer Jovian planets. Earth scientists have built up a hypothesis of planetary accretion that can explain the differences between the two groups. The *planetary accretion hypothesis*, a supplement to the nebula hypothesis we met earlier, states that planets assembled themselves from rocky, metallic, and icy debris 4.56 billion years ago, shortly after the Sun itself was formed.

The terrestrial planets: Mercury, Venus, Earth, and Mars

Earth scientists hypothesize that every terrestrial or rocky planet probably started out hot enough to melt either partially or completely, and it was during this period that its interior separated into layers with different chemical compositions, a process called *differentiation*. As we saw in Chapter 1, Earth differentiated into three layers: a relatively thin, low-density, rocky *crust*; a rocky, intermediate-density *mantle*; and a metallic, high-density *core*. Similar layers are present in Mercury, Venus, Mars, and our Moon, although they differ in size and composition. The most remarkable of these bodies is the innermost planet Mercury. Barely larger than the Moon, it has a core that makes up 42% of its volume and an estimated 80% of its mass (**FIGURE 17.12**). All the terrestrial planets have undergone intense impact cratering, but while the signs of this process are well hidden on Venus and Earth, they are particularly apparent on Mercury.

We do not know whether any of the terrestrial planets besides Earth have molten or partially molten cores. The molten outer core and the relatively rapid rotation of Earth give rise to Earth's strong magnetic field. Magnetic fields do exist on the other terrestrial planets, but they are much weaker than Earth's.

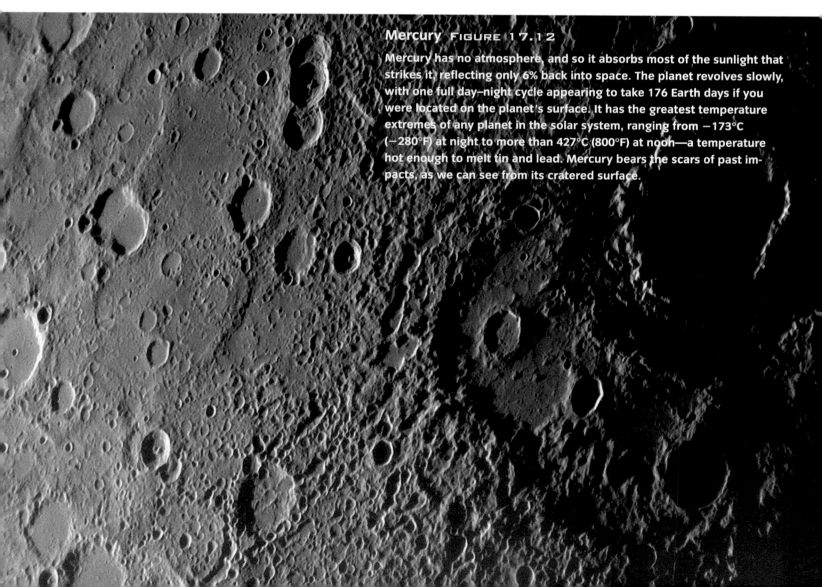

Mercury FIGURE 17.12

Mercury has no atmosphere, and so it absorbs most of the sunlight that strikes it, reflecting only 6% back into space. The planet revolves slowly, with one full day–night cycle appearing to take 176 Earth days if you were located on the planet's surface. It has the greatest temperature extremes of any planet in the solar system, ranging from −173°C (−280°F) at night to more than 427°C (800°F) at noon—a temperature hot enough to melt tin and lead. Mercury bears the scars of past impacts, as we can see from its cratered surface.

Venus is similar to Earth in size, mass, and distance from the Sun. However, its atmosphere is 97% carbon dioxide, and its surface reaches 475°C (900°F), making it inhospitable to life. The Magellan mission used radar to see through the dense clouds cloaking the planet and map the surface. The bright band running across the planet is a highland area of mountains and canyons. Many of the dark areas represent lava flows.

There are other important similarities among the terrestrial planets. All of them have experienced volcanic activity, Venus being a particularly good example (FIGURE 17.13). This means that they either have, or have had, an internal heat source, as we saw in Chapter 8. The volcanism is dominated by the formation of basalt, which forms most of the surface of Venus, Mars, and the dark-colored "seas" of Earth's Moon.

It is more difficult to determine whether each terrestrial planet has *plate tectonic* activity (described in Chapter 7). Simple observation of planets and their moons reveals that rocks on the surface of each planet fracture and deform as they do on Earth, which suggests that the planets have *lithospheres*. What little evidence we have suggests that *asthenospheres* and lithospheres probably are present in each terrestrial planet but that the asthenosphere of Earth is unusually near the surface and the lithosphere unusually thin. In fact, it is very likely that Earth is such a dynamic planet *because* its lithosphere is thin. The other terrestrial planets seem to have much thicker lithospheres and to be much less dynamic than Earth.

Finally, all terrestrial planets have lost their primordial atmospheres of hydrogen and helium. Earth, Mars, and Venus have new atmospheres that leaked from their interiors via volcanoes and were trapped by the planets' gravity. Mercury and the Moon are too small to have held onto the gases given off by volcanoes, so they lack atmospheres.

Given that the terrestrial planets are similar in composition and overall structure, why is life not present on Venus, or on Mars (highlighted in the *Amazing Places* feature at the end of the chapter)? The answer probably lies in two factors: (1) Earth's size and the fact that its interior is still hot leads to volcanism and the continual addition of new nutrients to the surface in the form of lava; (2) the distance of Earth from the Sun is such that H_2O can exist as water, ice, and water vapor. Not only do we need liquid water to live, but the fact that H_2O can exist as ice and water vapor plays a vital role in determining the character of our atmosphere and in creating habitable climates around the globe, as we saw in Chapter 14.

None of the gases of Jupiter's atmosphere have been able to escape its gravity, and storms are common. The wind systems create Jupiter's distinctive bright and dark bands. They are driven by heat generated in Jupiter's interior, in contrast to Earth's wind systems, which are driven by energy from the Sun.

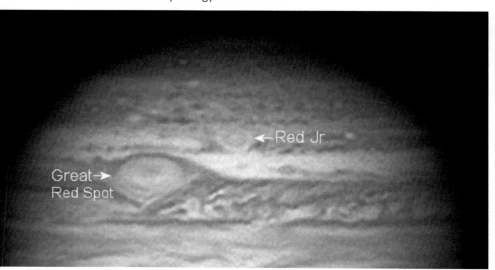

A The famed Great Red Spot on Jupiter, which is twice as wide as Earth, marks a storm that may have been raging for over 300 years. In 2006, an amateur astronomer noted a new storm, dubbed "Red Junior."

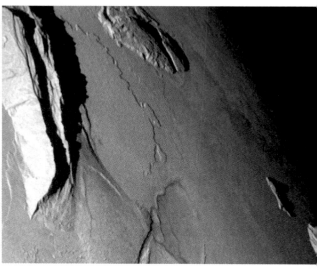

B NASA's Galileo spacecraft captured this dramatic image of mountains on one of Jupiter's moons, Io, in February 2000. Mongibello Mons, the jagged ridge at the left of the image, rises 23,000 ft above the plains of Io, higher than any mountain in North America.

The Jovian planets We cannot see anything that lies below the thick blankets of atmosphere—rich in hydrogen and helium—that cover the Jovian planets. But we can work out their internal structure based on remote-sensing measurements of the planets and their moons. Their cores are inferred to be rocky, surrounded by a thick layer of ice. Jupiter is a gas giant (FIGURE 17.14A) that is so large it has a mass that is two and a half times that of all the other planets in the solar system combined. If the planet had been 10 times larger, it would probably have developed into a small star—as described in the next section. Indeed, Jupiter lies at the center of its own satellite system, with more than 60 moons discovered so far (FIGURE 17.14B).

What an Earth Scientist Sees describes some recent observations of the next Jovian planet in line from the Sun, Saturn. The planet's famous ring system was first discovered by Galileo, with his primitive telescope. It was recently discovered that Jupiter, Neptune, and Uranus also have ring systems. The origin of these rings is still under debate. Saturn is similar in composition to Jupiter and also has a turbulent atmosphere.

Because pressures inside the Jovian planets must be enormous, we think that deep in their interiors hydrogen may be so tightly squeezed that it is condensed to a liquid. Still deeper inside Jupiter and Saturn, pressures equivalent to 30 million times the pressure at the surface of Earth are reached. Under such conditions, electrons and protons become less closely linked and hydrogen becomes metallic. In Jupiter, it is possible that pressures may even reach levels high enough for solid metallic hydrogen to form a sheath around the ice core.

Saturn's Rings

The Cassini probe took this spectacular view of Saturn in 2006 (A). The unique view of Saturn's rings was made possible because Saturn passed directly between the probe and the Sun, leaving the planet in silhouette. Viewing the rings through a visual and infrared mapping spectrometer reveals that the grains range from very small, like powdery snow on Earth, to larger grains, like more granular snow,

related to their distance from the planet (B). Cassini also confirmed that the grains are frozen water molecules, which supports the notion that Saturn may have a rocky core surrounded by a layer of ice.

The probe also glimpsed a tiny blue orb peeking out between Saturn's rings—Earth (C). Our world has an atmosphere that scatters blue light, and as a result it is the only planet that appears blue when viewed from deep space. Scientists looking for habitable planets orbiting neighboring suns will be encouraged if they spot a similar blue dot. But colors are only clues. A planet with an atmosphere might not appear blue. Mars has a thin red atmosphere. The blue light in the Martian skies is soaked up by iron-containing molecules on the surface that radiate red light.

Here's an interesting question:

CRITICAL THINKING

- What clues, in addition to color of the atmosphere, would you seek if you were a scientist looking for planets potentially habitable for humans?

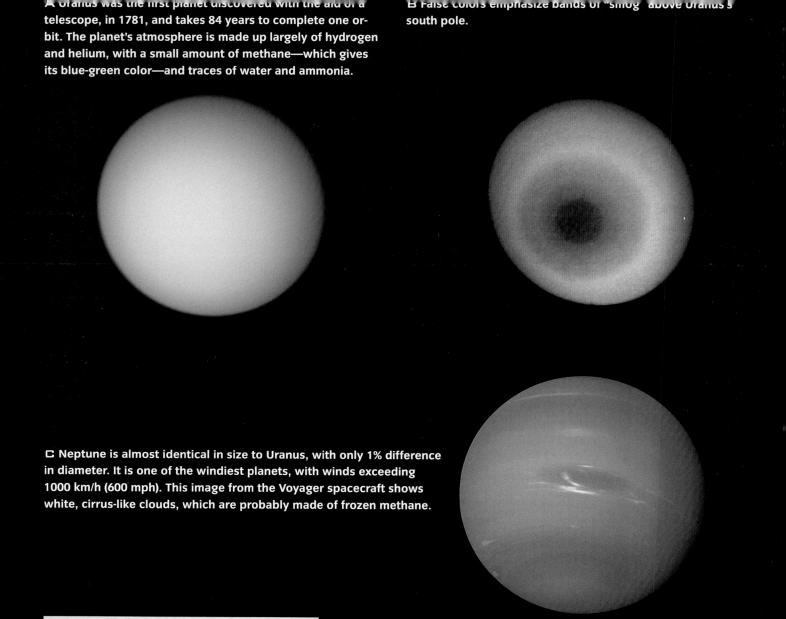

A Uranus was the first planet discovered with the aid of a telescope, in 1781, and takes 84 years to complete one orbit. The planet's atmosphere is made up largely of hydrogen and helium, with a small amount of methane—which gives its blue-green color—and traces of water and ammonia.

B False colors emphasize bands of "smog" above Uranus's south pole.

C Neptune is almost identical in size to Uranus, with only 1% difference in diameter. It is one of the windiest planets, with winds exceeding 1000 km/h (600 mph). This image from the Voyager spacecraft shows white, cirrus-like clouds, which are probably made of frozen methane.

Uranus and Neptune FIGURE 17.15

Neptune and Uranus are thought to be similar in composition to Jupiter and Saturn, although neither is large enough for pressures to reach the levels required for the formation of metallic hydrogen (FIGURE 17.15).

The dwarf planets

As we saw in the chapter opener, Pluto—once classified as the ninth planet in the solar system—was reclassified by the International Astronomical Union (IAU) as a "dwarf planet" in 2006. Like normal planets, dwarf planets orbit the Sun and have a large enough mass to pull their matter into a roughly spherical shape. However, unlike planets, they have not been able to clear their orbits of other objects of a similar size. There are currently two other objects that are classified as dwarf planets, Eris and Ceres. Ceres lies in the *asteroid belt*, which we will discuss in the next section.

Asteroids, meteorites, and the formation of the Moon The planetary accretion hypothesis also accounts for the existence of meteoroids and asteroids—small solar system bodies that are not massive enough to have pulled themselves into a round shape. Today's *meteoroids* are the debris that was never swept up to form a planet. Sometimes pieces of this debris happen to fall to Earth as **meteorites**. As such, meteorites are fascinating relics of the early days of the solar system (**FIGURE 17.16A**).

> **meteorite** A fragment of extraterrestrial material that falls to Earth.

Asteroids are rocky space objects that orbit the Sun, and can be as large as several hundred kilometers across. Most are concentrated in a belt between Mars and Jupiter, and take between three and six years to orbit the Sun. A few large asteroids pass close to Earth and the Moon, and probably caused the most recent impact craters on our planet.

It was only after the Moon landings in 1969–1972 that scientists began to grasp the great importance of violent collisions in the solar system's history. Every crater on the Moon, as far as we know, was formed by a meteoroid impact. (The astronauts looked for volcanic craters but found none.) Scientists recognize impact craters on Earth, too, although they are harder to find because erosion and other processes cover them up or erase them (**FIGURE 17.16B**).

Meteorites—messengers from space FIGURE 17.16

Global Locator

A This boy is examining a large iron meteorite in the American Museum of Natural History in New York City. Information gained from the study of such meteorites helps scientists interpret the distribution of chemicals in our solar system.

B Manicouagan Crater in Quebec was created about 210 million years ago by the impact of a much larger meteorite. The original crater, now marked by a ring lake, was 100 km in diameter.

Most planetary scientists now conclude that Earth collided with a planetary body roughly the same size as Mars about 4.5 billion years ago. That impact tilted Earth's axis of rotation at an angle of 23.5 degrees to the plane of its orbit around the Sun. As we saw earlier, this tilt explains why we have seasons. The impact must also have melted most of Earth's surface due to the tremendous amount of energy it released. (At the hyperspeeds typical of cosmic impacts, every ton of an impactor strikes Earth with an energy equivalent to 100 tons of dynamite.) The collision completely destroyed the other planet and blasted so much debris into orbit that for a while Earth had rings much denser than Saturn's. Eventually the rings of debris coalesced to form the Moon (**FIGURE 17.17**). This hypothesis explains the existence of a magma ocean early in lunar history (shown by rocks retrieved from the Moon). It also explains our Moon's relatively large size in contrast to other moons, which are many times smaller than their parent bodies, and accounts for certain chemical discrepancies and similarities between Earth and its Moon.

Such giant collisions were the inevitable final stage of planetary accretion, when most of the debris had been swept up and only larger objects remained. Signs of giant impacts abound. One such impact probably caused Uranus's axis to tip over on its side. Pluto's moon, Charon, shown in the photo at the beginning of the chapter, was probably created by an impact, because it is unusually large compared to Pluto. Perhaps Venus experienced a giant impact, too. Although it lacks a moon, it is the only planet that rotates clockwise on its axis, an effect that could have been produced by a large glancing blow that essentially turned the planet upside down.

Comets *Comets* have been dubbed "dirty snowballs" because they are balls of frozen gases (water, ammonia, methane, carbon dioxide, and carbon monoxide) and small pieces of rocky and metallic materials. They measure just a few kilometers or tens of kilometers across. The orbits of most comets are extremely elongated, taking hundreds of thousands of years to travel around the Sun and carrying them far beyond Pluto. However, a few comets, such as Halley's comet, have short-periods of less than 200 years, and regularly enter the vicinity of the planets. As comets approach the inner solar system, radiation from the Sun vaporizes their frozen gases, producing their distinctive glowing head, or *coma*. As they approach the Sun, some also develop a tail that can stream behind them for millions of kilometers (**FIGURE 17.18**).

Formation of the Moon FIGURE 17.17

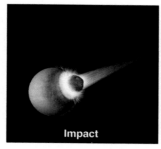

Impact

Some 4.5 billion years ago, the still-forming Earth runs into another growing planet, which scientists have dubbed Theia.

Impact + 8 hours

Theia is destroyed, and its remnants—along with a good chunk of Earth's mantle—are blasted into orbit around Earth. The off-center impact knocks Earth's axis of rotation askew.

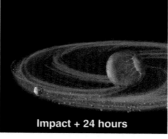

Impact + 24 hours

The debris spreads itself into a ring and begins to clump together.

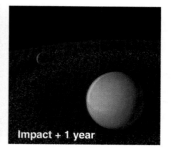

Impact + 1 year

The largest clump starts to attract other fragments and is well on its way toward becoming the Moon.

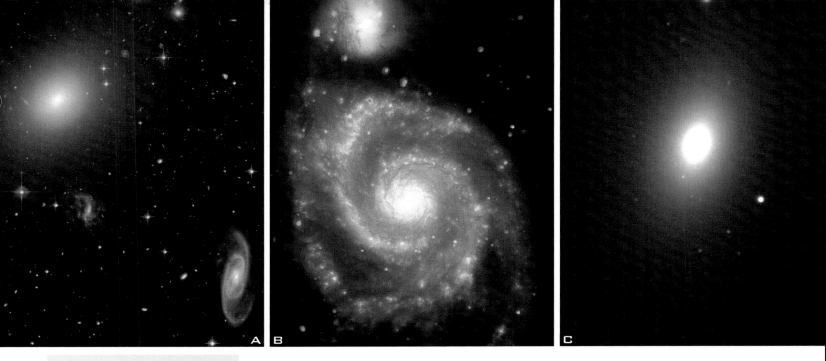

Galaxies FIGURE 17.24

A Taking long-exposure photographs over several hours, the Hubble Space Telescope reveals hundreds of galaxies in what would otherwise appear to be "empty" sky. These deep-field images provide evidence that the universe holds tens of billions of galaxies. Most are spiral galaxies, such as the Whirlpool Galaxy, shown in **B**. Another common type is the elliptical galaxy, such as this galaxy from the Virgo cluster, shown in **C**.

THE UNIVERSE TODAY

We now know that the universe contains countless galaxies (**FIGURE 17.24**). These galaxies fall into two main types: *spirals* and *elliptical galaxies*. About 75% of the brighter galaxies in the sky are spirals. Another 20% are elliptical galaxies, resembling cosmic footballs. The universe is also littered with irregular and dwarf galaxies, which could be the most common types of galaxies in the universe, but because they are faint, they are hard to detect. The Milky Way is an example of a spiral galaxy, as shown in Figure 17.20, with new stars forming in its spiral arms.

Star formation is usually an orderly process, but things are different at the heart of a small, but violent, minority of galaxies. About 10,000 *active galaxies* house *quasars* (quasi-stellar radio sources) at their cores. Quasars are relatively compact objects that spew vast amounts of energy into space. Astronomers do not yet understand the source of this energy, but many think that it could be due to a supermassive black hole lurking inside a quasar. As mentioned earlier, although we cannot see black holes directly, astronomers expect that huge amounts of energy would be released when matter falls into a black hole.

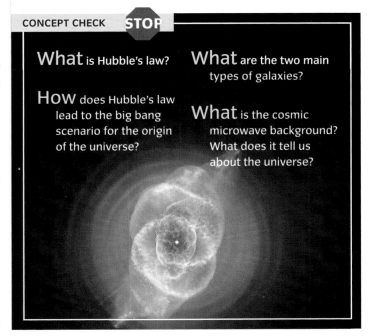

CONCEPT CHECK **STOP**

What is Hubble's law?

How does Hubble's law lead to the big bang scenario for the origin of the universe?

What are the two main types of galaxies?

What is the cosmic microwave background? What does it tell us about the universe?

Amazing Places: Mars

Humans have always been fascinated by Mars. For the past 40 years, NASA spacecraft have been increasing our understanding the red planet's climate and features.

A NASA's Hubble Space Telescope took this closeup of the red planet in August 2003, when it was just 34,648,840 mi (55,760,220 km) away, just 11 hours before the planet made its closest approach to Earth in 60,000 years. The planet's northern hemisphere is home to volcanoes that may have been active about 1 billion years ago. These volcanoes resurfaced the landscape, perhaps filling in many impact craters. In contrast, the southern hemisphere is pockmarked with ancient-impact craters.

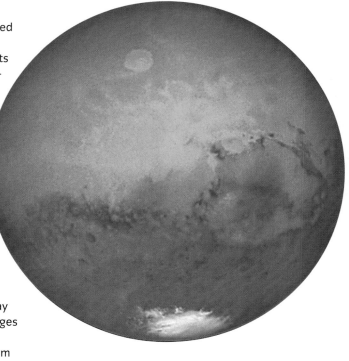

B In recent years, the Mars rovers, Spirit and Opportunity, and the Phoenix lander have been exploring the planet and sending home unique postcards showing its surface in unprecedented detail. In May 2008, Phoenix returned images of Mars' "quilted" surface. Similar polygon patterns form on icy ground in Earth's arctic regions.

C One of the most intriguing questions is whether water recently flowed on Mars. NASA's Mars Reconnaissance Orbiter captured images of gullies on Mars that could have been created by liquid water flowing on the planet within the last few million years. The gullies emanating from the rocky cliffs near the crater's rim (upper left) show meandering and braided patterns that are typical of channels carved by water. But recent analyses of mineral deposits on Mars suggest that any running water may have been too salty to support life.

SUMMARY

1 Astronomy and the Scientific Revolution

1. The concept of a **heliocentric universe**, with Earth rotating on its own axis and other planets revolving around the Sun, provides a simple explanation of the retrograde motion of *planets*.

2. The seasons occur because Earth's axis of rotation is tilted.

3. Kepler developed the following three laws of planetary motion: (a) The planets revolve around the Sun in elliptical orbits, (b) in such a way that a line drawn from a planet to the Sun sweeps out equal areas in equal amounts of time. (c) For any planet, the square of the orbital period in years is proportional to the cube of the planet's average distance from the Sun. Newton realized that the planets are held in orbit by their gravitational attraction to the Sun.

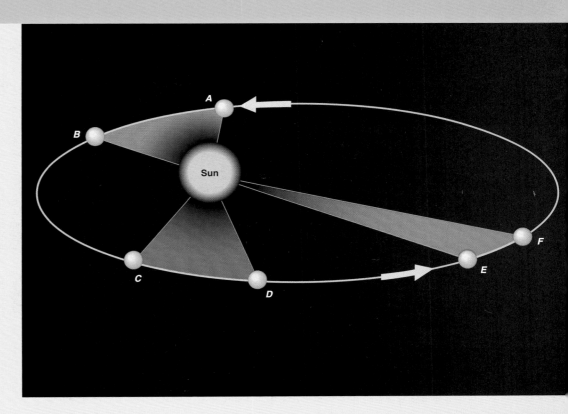

2 The Solar System

1. The *solar system* consists of the Sun, eight planets, many dwarf planets, numerous moons, and vast numbers of *asteroids*, millions of *comets*, and innumerable *meteoroids*, which can fall to Earth as **meteorites**.

2. The solar system was formed according to the **nebula hypothesis**, through the condensation of a solar nebula followed by *planetary accretion*, a process that was completed about 4.56 billion years ago.

3. Solar energy is generated in the Sun's core by the **proton–proton chain** of nuclear fusion reactions. Energy travels through the solar interior to the surface by *radiation* and *convection*. The Sun's outer layers are the *photosphere*, the *chromosphere*, and the corona.

4. The planets can be divided into two groups: the *terrestrial planets*, the four nearest the Sun, each a small, rocky mass with a high density; and the *Jovian planets*, the four outermost planets, each large and gassy.

5. The terrestrial planets have a metallic core, a mantle, and a crust.

6. The Jovian planets are shrouded by thick atmospheres that are rich in hydrogen and helium. Their cores are inferred to be rocky, like a terrestrial planet, surrounded by a thick layer of ice; above the ice is liquid hydrogen, which grades outward to the hydrogen-rich atmosphere.

3 Stars and Stellar Evolution

1. Stars are characterized on a **Hertzsprung-Russell diagram** by their surface temperature and luminosity. A star's fate is determined by its initial mass. Low-mass stars will evolve as *brown dwarfs*. Stars with a mass similar to our Sun will become *red giants* and then *white dwarfs*. Higher-mass stars will explode as **supernovae**, and their cores may become *neutron stars* or *black holes*.

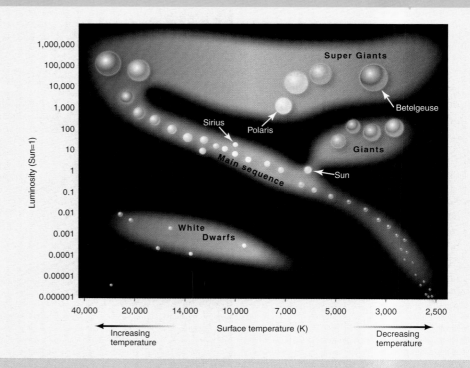

4 The Universe and How It Came to Be

1. The Milky Way is one of countless *galaxies*. These galaxies are receding from one another, and according to **Hubble's law**, the farther the galaxy is from Earth, the faster it is receding.

2. The **big bang theory** states that the universe began at a specific moment in time and has been expanding ever since. The recession of the galaxies, the discovery of the **cosmic microwave background radiation**, and the abundance of light elements in the universe are evidence for the big bang.

3. There are two main types of galaxies: *spirals* and *elliptical galaxies*. The Milky Way is a spiral galaxy.

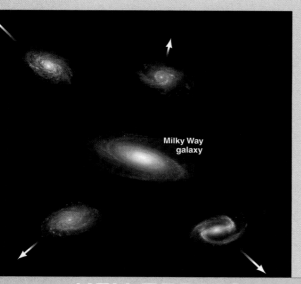

KEY TERMS

CRITICAL AND CREATIVE THINKING QUESTIONS

1. NASA scientists have focused more attention on Mars than any other planet in the solar system. One of the reasons for doing so is that Mars may once have had an environment on which life might have started. If you had a chance to seek evidence of life elsewhere on the planets or moons of the solar system, what would you choose as targets?

2. Earth scientists have a hypothesis that the last of the dinosaurs were driven to extinction at the end of the Cretaceous Period by the impact of a large meteorite. What might the consequences be if an equally large meteorite were to strike Earth today? Consider two possibilities: first, that the site of the impact be in the center of the Atlantic Ocean; and second, that the impact be in the fertile farming region of the midwestern United States.

3. Since the time of Aristotle, our view of our place in the universe has changed dramatically. Astronomical observations have played a major role in changing our view. How do you think our view of our place in the universe would change if scientists discovered evidence of intelligent life on a planet orbiting some other star?

4. The Hubble Space telescope has gathered exceptional evidence of stars and other things in our Milky Way galaxy, and in galaxies in the universe beyond. Imagine you are a member of Congress meeting with scientists who seek funds for an even more powerful eye-in-the sky. What scientific justifications would you ask of the scientists in order to decide whether to support or deny their request?

5. Discussions are now being held in both space science and political circles about a manned space mission to Mars. The mission would be very dangerous for astronauts and the expense would be great. How would you justify such a mission? If you don't agree with a manned mission, what would your objectives be for future unmanned missions?

What is happening in this picture ?

This Hubble image shows a possible exoplanet orbiting a star.

What features would make an exoplanet hospitable for life?

If it were possible to send an unmanned spaceship to a nearby star in order to inspect the planets in orbit around the star, what kind of measurements would you advise the space scientists to try to obtain as the spaceship flies past each planet?

1. Who first proposed a heliocentric model of the universe?
   a. Aristarchus
   c. Copernicus
   b. Aristotle
   d. Kepler

2. Which of the following was *not* discovered by Kepler?
   a. Planets' orbits around the Sun are elliptical due to the force of gravity.
   b. Planets change speed as they move along their orbits.
   c. Planets sweep out equal areas along their orbits, in equal times.
   d. Planets move such that the square of their period in years is proportional to the cube of the planet's average distance from the Sun.

3. What feature, shown here, did Galileo observe, supporting the heliocentric universe?
   a. phases of the Moon
   c. partial eclipse of the Moon
   b. phases of Venus
   d. transit of Venus

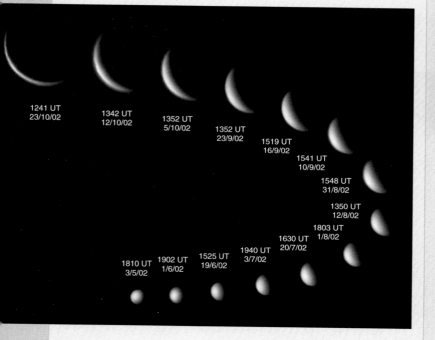

4. Label the following regions on this diagram of the Sun: (a) chromosphere, (b) corona, (c) core, (d) convection zone, (e) photosphere, (f) radiation zone.

5. The Sun derives its energy from _____.
   a. hydrogen fission
   c. helium fission
   b. hydrogen fusion
   d. helium fusion

6. Which of the following features is not shared by all terrestrial planets?
   a. an atmosphere
   c. a magnetic field
   b. a core
   d. a crust

7. The pressure inside Jupiter's core is so high that it is possible that _____.
   a. hydrogen gas could condense to liquid hydrogen
   b. hydrogen atoms could fuse to form helium atoms
   c. hydrogen becomes metallic
   d. (a) and (b)
   e. (a) and (c)
   f. (b) and (c)

8. The Jovian planets—Jupiter, Saturn, Uranus, and Neptune—all share which of the following features?
   a. molten cores
   b. metallic hydrogen sheaths surrounding their cores
   c. thick atmospheres
   d. all of the above

9. The asteroid belt lies between _____ and _____.
   a. Earth/Mars
   c. Jupiter/Saturn
   b. Mars/Jupiter
   d. Saturn/Uranus

10. Planetary _____ accounts for the existence of asteroids and meteoroids.
    a. condensation
    c. accretion
    b. aggregation
    d. clustering

11. Comets are thought to originate in _____.
    a. the Oort cloud
    d. (a) and (b)
    b. the Kuiper belt
    e. (a) and (c)
    c. the asteroid belt
    f. (b) and (c)

12. The Moon is thought to have been formed when a _____-sized object collided with Earth.
    a. Mercury
    c. Mars
    b. Venus
    d. Uranus

13. Why might the exoplanet Gliese 581c not be habitable?
    a. It probably does not have a rocky composition.
    b. Its atmosphere is probably poisonous.
    c. It is probably too cold.
    d. It is probably undergoing a runaway greenhouse effect.

14. What objects observed by Hubble enabled him to work out the distance to neighboring galaxies?
    a. quasars
    c. Cepheid variables
    b. dwarf galaxies
    d. supernovae

15. Which of the following does not serve as evidence for the big bang theory of the origin of the universe?
    a. the amount of light elements in the universe
    b. the direction of rotation of the Sun and the planets (excluding Venus)
    c. the recession of galaxies
    d. the cosmic microwave background

Appendix A
Periodic Table of the Elements

The periodic table lists the known **chemical elements**, the basic units of matter. The elements in the table are arranged left-to-right in rows in order of their **atomic number**, the number of protons in the nucleus. Each horizontal row, numbered from 1 to 7, is a **period**. All elements in a given period have the same number of electron shells as their period number. For example, each atom of hydrogen or helium has one electron shell, while each atom of potassium or calcium has four electron shells. The elements in each column, or **group**, share chemical properties. For example, the ele-

ments in column IA are very chemically reactive, whereas the elements in column VIIIA have full electron shells and thus are chemically inert.

Scientists now recognize up to 118 different elements; 92 occur naturally on Earth, and the rest (with the exception of element 117) have been produced synthetically using particle accelerators. Elements are designated by **chemical symbols**, which are the first one or two letters of the element's name in English, Latin, or another language.

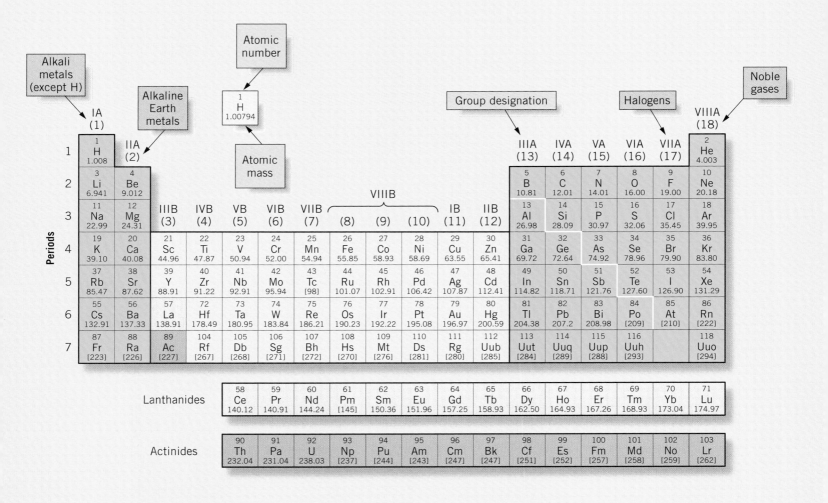

Appendix B
Units and Their Conversions

Commonly Used Units of Measure

Length

Metric Measure

1 kilometer (km)	= 1000 meters (m)
1 meter (m)	= 100 centimeters (cm)
1 centimeter (cm)	= 10 millimeters (mm)
1 millimeter (mm)	= 1000 micrometers (μm) (formerly called microns)
1 micrometer (μm)	= 0.001 millimeter (mm)
1 angstrom (Å)	= 10^{-8} centimeters (cm)

Nonmetric Measure

1 mile (mi)	= 5280 feet (ft) = 1760 yards (yd)
1 yard (yd)	= 3 feet (ft)
1 fathom (fath)	= 6 feet (ft)

Conversions

1 kilometer (km)	= 0.6214 mile (mi)
1 meter (m)	= 1.094 yards (yd)
	= 3.281 feet (ft)
1 centimeter (cm)	= 0.3937 inch (in)
1 millimeter (mm)	= 0.0394 inch (in)
1 mile (mi)	= 1.609 kilometers (km)
1 yard (yd)	= 0.9144 meter (m)
1 foot (ft)	= 0.3048 meter (m)
1 inch (in)	= 2.54 centimeters (cm)
1 inch (in)	= 25.4 millimeters (mm)
1 fathom (fath)	= 1.8288 meters (m)

Area

Metric Measure

1 square kilometer (km^2)	= 1,000,000 square meters (m^2)
	= 100 hectares (ha)
1 square meter (m^2)	= 10,000 square centimeters (cm^2)
1 hectare (ha)	= 10,000 square meters (m^2)

Nonmetric Measure

1 square mile (mi^2)	= 640 acres (ac)
1 acre (ac)	= 4840 square yards (yd^2)
1 square foot (ft^2)	= 144 square inches (in^2)

Conversions

1 square kilometer (km^2)	= 0.386 square mile (mi^2)
1 hectare (ha)	= 2.471 acres (ac)
1 square meter (m^2)	= 1.196 square yards (yd^2)
	= 10.764 square feet (ft^2)
1 square centimeter (cm^2)	= 0.155 square inch (in^2)
1 square mile (mi^2)	= 2.59 square kilometers (km^2)
1 acre (ac)	= 0.4047 hectare (ha)
1 square yard (yd^2)	= 0.836 square meter (m^2)
1 square foot (ft^2)	= 0.0929 square meter (m^2)
1 square inch (in^2)	= 6.4516 square centimeter (cm^2)

Volume

Metric Measure

1 cubic meter (m^3)	= 1,000,000 cubic centimeters (cm^3)
1 liter (l)	= 1000 milliliters (ml)
	= 0.001 cubic meter (m^3)
1 centiliter (cl)	= 10 milliliters (ml)
1 milliliter (ml)	= 1 cubic centimeter (cm^3)

Nonmetric Measure

1 cubic yard (yd^3)	= 27 cubic feet (ft^3)
1 cubic foot (ft^3)	= 1728 cubic inches (in^3)
1 barrel (oil) (bbl)	= 42 gallons (U.S.) (gal)

Conversions

1 cubic kilometer (km^3)	= 0.24 cubic miles (mi^3)
1 cubic meter (m^3)	= 264.2 gallons (U.S.) (gal)
	= 35.314 cubic feet (ft^3)
1 liter (l)	= 1.057 quarts (U.S.) (qt)
	= 33.815 ounces (U.S. fluid) (fl. oz.)
1 cubic centimeter (cm^3)	= 0.0610 cubic inch (in^3)
1 cubic mile (mi^3)	= 4.168 cubic kilometers (km^3)
1 acre-foot (ac-ft)	= 1233.46 cubic meters (m^3)
1 cubic yard (yd^3)	= 0.7646 cubic meter (m^3)
1 cubic foot (ft^3)	= 0.0283 cubic meter (m^3)
1 cubic inch (in^3)	= 16.39 cubic centimeters (cm^3)
1 gallon (gal)	= 3.784 liters (l)

Mass

Metric Measure

1000 kilograms (kg)	= 1 metric ton (also called a *tonne*) (m.t)
1 kilogram (kg)	= 1000 grams (g)

Nonmetric Measure

1 short ton (sh.t)	= 2000 pounds (lb)
1 long ton (l.t)	= 2240 pounds (lb)
1 pound (avoirdupois) (lb)	= 16 ounces (avoirdupois) (oz) = 7000 grains (gr)
1 ounce (avoirdupois) (oz)	= 437.5 grains (gr)
1 pound (Troy) (Tr. lb)	= 12 ounces (Troy) (Tr. oz)
1 ounce (Troy) (Tr. oz)	= 20 pennyweight (dwt)

Conversions

1 metric ton (m.t)	= 2205 pounds (avoirdupois) (lb)
1 kilogram (kg)	= 2.205 pounds (avoirdupois) (lb)
1 gram (g)	= 0.03527 ounce (avoirdupois) (oz) = 0.03215 ounce (Troy) (Tr. oz) = 15,432 grains (gr)
1 pound (lb)	= 0.4536 kilogram (kg)
1 ounce (avoirdupois) (oz)	= 28.35 grams (g)
1 ounce (avoirdupois) (oz)	= 1.097 ounces (Troy) (Tr. oz)

Pressure

Metric Measure

1 pascal (Pa)	= 1 newton/square meter (N/m^2)
1 kilogram-force square centimeter (kg/cm^2 or kgf/cm^2)	= 1 technical atmosphere (at) = 98,067 Pa = 0.98067 bar
1 bar	= 10^5 pascals (Pa) = 1.02 kilogram-force/square centimeter (kgf/cm^2 or at)

Nonmetric Measure

1 atmosphere (atm)	= 1.01325 bar = 14.696 lb/in^2 (psi)
1 pound per square inch (lb/in^2 or psi)	= 68.046×10^{-3} atmospheres (atm)

Conversions

1 kilogram-force/ square centimeter (kgf/cm^2 or at)	= 0.96784 atmosphere (atm) = 14.2233 pounds/square inch (lb/in^2 or psi) = 0.98067 bar = 98,067 Pa
1 bar	= 0.98692 atmosphere (atm) = 10^5 pascals (Pa) = 1.02 kilogram-force/square centimeter (kgf/cm^2)
1 Pa = 10^{-5} bar	= 10.197×10^{-6} kgf/cm^2 (or at) = 9.8692×10^{-6} atm = 145.04×10^{-6} lb/in^2 (or psi)
1 atm	= 101,325 Pa = 1.01325 bar

Temperature

Metric Measure

0 degrees Celsius (°C)	= freezing point of water at sea level
100 degrees Celsius (°C)	= boiling point of water at sea level
0 degrees Kelvin (K)	= −273.15°C = absolute zero
1°C	= 1K (temperature increments)
273.15K	= 0.0° C

Nonmetric Measure

Fahrenheit (°F)	= (K · 9/5) − 459.67
Fahrenheit (°F)	= (°C · 9/5) + 32

Conversions

degrees Kelvin (K)	= °C + 273.1
degrees Celcius (°C)	= K − 273.15
degrees Fahrenheit (°F)	= (°C · 9/5) + 32
degrees Celcius(°C)	= (°F − 32) · 5/9

Appendix C
Answers to Self-Tests

CHAPTER 1
1. c;   **2.** d;   **3.** See Figure 1.5;   **4.** d;   **5.** b;   **6.** b;   **7.** d;   **8.** See Figure 1.9;   **9.** c;   **10.** b;   **11.** See Figure 1.10;   **12.** c;   **13.** d; **14.** e;   **15.** e

CHAPTER 2
1. d;   **2.** See Figure 2.1;   **3.** b;   **4.** c;   **5.** e;   **6.** d;   **7.** a;   **8.** b; **9.** a;   **10.** c;   **11.** e;   **12.** b;   **13.** a;   **14.** See Table 2.2;   **15.** d

CHAPTER 3
1. See Figure 3.2;   **2.** c;   **3.** A = phaneritic-texture plutonic rock B = porphyritic-texture volcanic rock;   **4.** b;   **5.** d;   **6.** a;   **7.** b; **8.** b;   **9.** a;   **10.** e;   **11.** d;   **12.** d;   **13.** c;   **14.** c;   **15.** c

CHAPTER 4
1. b ;   **2.** c;   **3.** d;   **4.** e;   **5.** c;   **6.** a;   **7.** See Figure 4.11;   **8.** d; **9.** d;   **10.** b;   **11.** d;   **12.** c;   **13.** b;   **14.** See Figure 4.17;   **15.** c

CHAPTER 5
1. See Figure 5.1;   **2.** a;   **3.** d;   **4.** c;   **5.** d;   **6.** b;   **7.** d;   **8.** d; **9.** b;   **10.** c;   **11.** d;   **12.** c;   **13.** b;   **14.** b;   **15.** See Figure 5.15.

CHAPTER 6
1. b;   **2.** c;   **3.** d;   **4.** d;   **5.** See Figure 6.7;   **6.** b;   **7.** d;   **8.** b; **9.** See Figure 6.13;   **10.** a;   **11.** b;   **12.** c;   **13.** c;   **14.** b;   **15.** c

CHAPTER 7
1. a;   **2.** c;   **3.** d;   **4.** c
5.

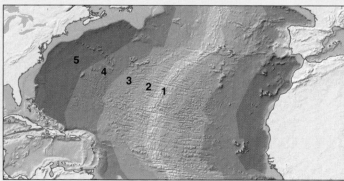

6. a
7.

Approximate ages of islands in millions of years

Kauai, 3–5.5 → A
B
Oahu, 2.25–3.25 → C
D
Molokai, 1.25–2 →
Maui, 0.5–1 → E

Hawaii, present–0.8

8. b;   **9.** c;   **10.** a;   **11.** d;   **12.** c;   **13.** See Figure 7.12;   **14.** See Figure 7.12;   **15.** c

CHAPTER 8
1. c;   **2.** b;   **3.** d;   **4.** b;   **5.** a;   **6.** a;   **7.** See Figure 8.12;   **8.** d; **9.** d;   **10.** e;   **11.** a;   **12.** d;   **13.** a;   **14.** b;   **15.** b

CHAPTER 9
1. a;   **2.** b;   **3.** b;   **4.** b;   **5.** c;   **6.** a;   **7.** d;   **8.** b;   **9.** b;   **10.** c; **11.** c;   **12.** c;   **13.** See Figure 9.20;   **14.** b;   **15.** c

CHAPTER 10
1. b;   **2.** c;   **3.** c;   **4.** c;   **5.** c;   **6.** See Figure 10.3;   **7.** d;   **8.** b; **9.** b;   **10.** b;   **11.** See Figure 10.9;   **12.** b;   **13.** a;   **14.** b;   **15.** d

CHAPTER 11
1. a;   **2.** See Figure 11.2B;   **3.** c;   **4.** c;   **5.** See Figure 11.8; **6.** e;   **7.** c;   **8.** b;   **9.** e;   **10.** e;   **11.** b;   **12.** d;   **13.** d;   **14.** a; **15.** d

CHAPTER 12
1. c;   **2.** a;   **3.** d ;   **4.** c;   **5.** a;   **6.** c;   **7.** d;   **8.** b;   **9.** b;   **10.** d; **11.** a;   **12.** b;   **13.** b;   **14.** a;   **15.** b

CHAPTER 13
1. e;   **2.** See Figure 13.3;   **3.** a;   **4.** a;   **5.** See Figure in *What an Earth Scientist Sees*, page 401;   **6.** d;   **7.** b;   **8.** d;   **9.** See Figure 13.11;   **10.** c;   **11.** b;   **12.** b;   **13.** a;   **14.** c;   **15.** a

CHAPTER 14
1. See Figure 14.1B;   **2.** a;   **3.** c;   **4.** See Figure 14.2;   **5.** c; **6.** b;   **7.** b;   **8.** b;   **9.** c;   **10.** a;   **11.** c;   **12.** a;   **13.** b;   **14.** a; **15.** a

CHAPTER 15
1. c;   **2.** c;   **3.** a;   **4.** c;   **5.** b;   **6.** See Figure 15.10;   **7.** d; **8.** See Figure 15.12;   **9.** a;   **10.** b;   **11.** c;   **12.** a;   **13.** c;   **14.** b; **15.** a

CHAPTER 16
1. a;   **2.** c;   **3.** d;   **4.** c;   **5.** e;   **6.** c;   **7.** b;   **8.** d;   **9.** See Figure 16.7;   **10.** b;   **11.** c;   **12.** c;   **13.** c;   **14.** e;   **15.** d

CHAPTER 17
1. a;   **2.** a;   **3.** b;   **4.** See Figure 17.11B;   **5.** b;   **6.** a;   **7.** c; **8.** c;   **9.** b;   **10.** c;   **11.** d;   **12.** c;   **13.** d;   **14.** c;   **15.** b

abrasion Wind erosion in which airborne particles chip small fragments off rocks that protrude above the surface.

adiabatic principle A principle of science that states that a gas cools as it expands and warms as it is compressed, provided that no heat flows into or out of the gas during the process.

air mass An extensive body of air in which temperature and moisture characteristics are fairly uniform over a large area.

alluvium Unconsolidated sediment deposited by a stream.

angiosperm A flowering, or seed-enclosed, plant.

anthropogenic Produced by human activities.

anticyclone A center of high atmospheric pressure.

aquiclude A layer of impermeable rock.

aquifer A body of rock or regolith that is water-saturated, porous, and permeable.

asthenosphere A layer of weak, ductile rock in the mantle that is close to melting but not actually molten.

atmospheric pressure Pressure exerted by the atmosphere at Earth's surface because of the force of gravity acting on the overlying column of air.

atom The smallest individual particle that retains the distinctive chemical properties of an element.

barometer An instrument that measures atmospheric pressure.

barrier island A long, narrow, sandy island lying offshore and parallel to a lowland coast.

batholith A large, irregularly shaped pluton that cuts across the layering of the rock into which it intrudes.

beach drift The movement of particles along a beach as they are driven up and down the beach slope by wave action.

beach Wave-washed sediment along a coast.

bed load Sediment that is moved along the bottom of a stream.

bedding The layered arrangement of strata in a body of sediment or sedimentary rock.

big bang theory The theory that the universe began to expand after an explosion of concentrated matter and energy.

biogenic sediment Sediment that is composed primarily of plant and animal remains, or precipitates as a result of biologic processes.

body wave A seismic wave that travels through Earth's interior.

bond The force that holds the atoms together in a chemical compound.

Bowen's reaction series The order in which minerals crystallize from, and subsequently react with, a cooling magma.

brittle deformation A permanent change in shape or volume, in which a material breaks or cracks.

cave and **cavern** Underground open space; a cavern is a system of connected caves.

cell The basic structural and functional unit of life; a complex grouping of chemical compounds enclosed in a porous membrane.

channel The clearly defined natural passageway through which a stream flows.

chemical sediment Sediment formed by the precipitation of minerals dissolved in lake, river, or seawater.

chemical weathering The decomposition of rocks and minerals by chemical and biochemical reactions.

clastic sediment Sediment formed from fragmented rock and mineral debris produced by weathering and erosion.

clay A family of hydrous aluminosilicate minerals. The term is also used for tiny mineral particles of any kind that have physical properties like those of the clay minerals.

cleavage The tendency of a mineral to break in preferred directions along bright, reflective plane surfaces.

climate The annual cycle of prevailing weather conditions at a given place based on statistics recorded over a long period.

coal A combustible rock formed from the lithification of plant-rich sediment.

compound A combination of atoms of one or more elements in a specific ratio.

compression A stress that acts in a direction perpendicular and *toward* a surface.

compressional wave A seismic body wave consisting of alternate pulses of compression and expansion in the direction of travel; P wave or primary wave.

condensation The process by which water changes from vapor into a liquid.

conduction The process by which heat moves through a solid body without deforming it.

conglomerate A clastic sedimentary rock with large fragments in a finer-grained matrix.

continental crust The older, thicker, and less dense part of Earth's crust; the bulk of Earth's land masses.

continental drift The slow, lateral movement of continents across Earth's surface.

convection A form of heat transfer in which hot material circulates from hotter to colder regions, loses its heat, and then repeats the cycle.

convection loop A circuit of moving fluid, such as air or water, created by unequal heating of the fluid.

convectional precipitation Precipitation that is induced when warm, moist air is heated at the ground surface, rises, cools, and condenses to form water droplets, raindrops, and, eventually, rainfall.

convergent margin A boundary along which two plates move toward one another.

core Earth's innermost compositional layer, where the magnetic field is generated and much geothermal energy resides.

Coriolis effect The effect of Earth's rotation, which acts like a force to deflect a moving object on the surface to the right in the Northern Hemisphere and to the left in the Southern Hemisphere.

correlation A method of equating the ages of strata that come from two or more different places.

cosmic microwave background (CMB) radiation Radiation left over from the big bang, detected in the microwave portion of the spectrum with a temperature of only 2.7 K.

creep The imperceptibly slow downslope granular flow of regolith.

crust The outermost compositional layer of the solid Earth; part of the lithosphere.

cryosphere The perennially frozen part of the hydrosphere.

crystal structure An arrangement of atoms or molecules into a regular geometric lattice. Materials that possess a crystal structure are said to be crystalline.

crystallization The process whereby mineral grains form and grow in a cooling magma (or lava).

cyclone A center of low atmospheric pressure.

deflation Wind erosion in which loose particles of sand and dust are removed by the wind, leaving coarser particles behind.

delta Sediment deposit built by a stream entering a body of standing water.

density The mass of material per unit volume.

deposition (1) The laying down of sediment.

deposition (2) The process by which water changes from a vapor into a solid.

desert An arid land that receives less than 250 mm of rainfall or snow equivalent per year, and is sparsely vegetated unless it is irrigated.

desertification Invasion of desert conditions into nondesert areas.

discharge (1) The amount of water passing by a point on the channel's bank during a unit of time.

discharge (2) The process by which subsurface water leaves the saturated zone and becomes surface water.

dissolution The separation of a material into ions in solution by a solvent, such as water or acid.

divergent margin A boundary along which two plates move apart from one another.

divide A topographic high that separates adjacent drainage basins.

DNA Deoxyribonucleic acid; a double-chain biopolymer that contains all the genetic information needed for organisms to grow and reproduce.

domain The broadest taxonomic category of living organisms; biologists today recognize three domains: bacteria, *Archaea*, and eukaryotes.

drainage basin The total area from which water flows into a stream.

ductile deformation A permanent but gradual change in the shape or volume of a material, caused by flowing or bending.

dune A hill or ridge of sand deposited by the wind.

Earth science The scientific study of all aspects of Earth.

Earth system science The study of Earth as a closed system composed of interacting open systems and how the open systems may be changed as a result of human activities.

earthquake A sudden motion in Earth caused by the abrupt release of slowly accumulated energy.

El Niño An episodic cessation of the normal upwelling of cold water off the coast of Peru.

elastic deformation A temporary change in the shape or volume from which a material rebounds after the deforming stress is removed.

elastic rebound theory The theory that continuing stress along a fault results in a buildup of elastic energy in the rocks, which is abruptly released when an earthquake occurs.

element The most fundamental substance into which matter can be separated by chemical means.

epicenter The point on Earth's surface that is directly above an earthquake's focus.

erosion The wearing away of bedrock and transport of loosened particles by a fluid, such as water.

eukaryotes Organisms composed of eukaryotic cells—that is, cells that have a well-defined nucleus and organelles.

evaporation The process by which water changes from a liquid to a vapor.

evaporite A rock formed by evaporation of lake water or seawater, followed by lithification of the resulting salt deposit.

evolution The theory that life on Earth has developed gradually, from one or a few simple organisms to more complex organisms.

exoplanet Short for *extrasolar planet*, a planet outside our solar system.

fault A fracture in the lithosphere along which movement has occurred.

feedback mechanisms Reactions that enhance (positive) or retard (negative) change in an open system.

flood An event in which a water body overflows its banks.

floodplain The relatively flat valley floor adjacent to a stream channel, which is inundated when the stream overflows its banks.

flow Any mass-wasting process that involves a flowing motion of regolith containing water and/or air within its pores.

focus The location where rupture commences and an earthquake's energy is first released.

foliation A planar arrangement of textural features in a metamorphic rock, which give the rock a layered or finely banded appearance.

fossil Remains of an organism from a past age, embedded and preserved in Earth's crust.

fractional crystallization Separation of crystals from liquids during crystallization.

fractional melt A mixture of molten and solid rock.

fractionation   Separation of melted materials from the remaining solid matter during the course of melting.

front   The surface of contact between two dissimilar air masses.

geocentric   A model of the universe in which a stationary Earth is at the center and all other celestial bodies revolve around it.

geologic column   The succession of strata, fitted together in relative chronological order.

geostrophic wind   Wind at high levels above Earth's surface blowing parallel with a system of straight, parallel isobars.

geothermal gradient   The rate at which temperature increases with depth below Earth's surface.

glaciation   The covering of large land areas by glaciers or ice sheets.

glacier   A semipermanent or perennially frozen body of ice, consisting largely of recrystallized snow, that moves under the pull of gravity.

gneiss   A coarse-grained, high-grade metamorphic rock.

gradient   The steepness of a stream channel.

greenhouse effect   The accumulation of heat in the lower atmosphere through the absorption of longwave radiation from Earth's surface.

groundwater   Subsurface water contained in pore spaces in regolith and bedrock.

gymnosperm   A naked-seed plant.

habit   The distinctive shape of a particular mineral.

half-life   The time needed for half the parent atoms of a radioactive substance to decay into daughter atoms.

hardness   A mineral's resistance to scratching.

heliocentric   A model of the universe in which a stationary Sun is at the center and the planets revolve around it.

Hertzsprung-Russell (H-R) diagram   A plot of stars according to their luminosity and surface temperature.

Hubble's law   The observed relationship that the velocity of recession of a galaxy is proportional to its distance.

humidity   The amount of water vapor in the air.

humus   Partially decayed organic matter in soil.

hydrologic cycle   A model that describes the movement of water through the reservoirs of the Earth system; the *water cycle*.

hypothesis   A plausible, but yet to be proved, explanation for how something happens.

ice sheet   The largest type of glacier on Earth, a continent-sized mass of ice that covers all or nearly all the land within its margins.

igneous rocks   Rocks that form by cooling and solidification of molten rock.

infiltration   The process by which water works its way into the ground through small openings in the soil.

insolation   The flow of incoming solar radiation intercepted by an exposed surface, assuming a uniformly spherical Earth with no atmosphere.

intertropical convergence zone (ITCZ)   A zone of convergence of air masses along the equatorial trough.

isobars   Lines on a map drawn through all points having the same atmospheric pressure.

isotopes   Atoms with the same atomic number and different mass numbers.

jet streams   High-speed air flows in narrow bands within the upper-air westerlies and along certain other global latitude zones at high levels.

joint   A fracture in a rock, along which no appreciable movement has occurred.

kingdom   The second-broadest taxonomic category. There are six recognized kingdoms, including animals and plants.

lapse rate   The rate at which air temperature decreases with increasing altitude.

latent heat   Heat that is absorbed and stored in a gas or liquid during the processes of evaporation or condensation, melting or freezing, sublimation or deposition.

lava   Molten rock that reaches Earth's surface.

limestone   A sedimentary rock that consists primarily of the mineral calcite.

lithification   The group of processes by which sediment is transformed into sedimentary rock.

lithosphere   Earth's outermost rocky layer, comprising the crust and the uppermost part of the mantle.

littoral drift   Transport of sediment parallel to the shoreline by the combined action of beach drift and longshore current transport.

load   The suspended and dissolved sediment carried by a stream.

longshore current   A current within the surf zone that flows parallel to the coast.

low-latitude rainforest   Dense evergreen forest of low-latitude equatorial and tropical zones with abundant rainfall.

luster   The quality and intensity of light that reflects from a mineral.

magma   Molten rock, with any suspended mineral grains and dissolved gases, that forms when melting occurs in the crust or mantle.

magmatic differentiation   The formation of many different kinds of igneous rock from a single magma.

magnetic reversal   The period of time in which Earth's magnetic polarity reverses itself.

mantle   The middle compositional layer of Earth, between the crust and the core.

marble   The product of metamorphism formed by the recrystallization of limestone.

mass extinction   A catastrophic episode in which a large fraction of living species become extinct within a geologically short time.

mass wasting   The downslope movement of regolith and/or bedrock masses due to the pull of gravity.

mechanical weathering   The breakdown of rock into solid fragments by physical processes that do not change the rock's chemical composition.

metamorphic rocks   Rocks that have been altered by exposure to high temperature, high pressure, or both.

metamorphism The mineralogical, textural, chemical, and structural changes that occur in rocks as a result of exposure to elevated temperatures and pressures.

meteorite A fragment of extraterrestrial material that falls to Earth.

Milankovitch cycles Climate cycles that occur over tens to hundreds of thousands of years because of changes in Earth's orbit and tilt.

mineral A naturally formed, solid, inorganic substance with a characteristic crystal structure and a specific chemical composition.

molecule The smallest chemical unit that has all the properties of a particular compound.

moment magnitude A measure of earthquake strength based on the rupture size, rock properties, and amount of displacement on the fault surface.

moraine A ridge or pile of debris that has been, or is being, transported by a glacier.

natural resources Useful materials obtained from the lithosphere, atmosphere, hydrosphere, or biosphere.

natural selection The process by which individuals that are well adapted to their environment have a survival advantage, and pass on their favorable characteristics to their offspring.

nebula hypothesis A model that explains the formation of the solar system from a large cloud of gas and dust floating in space 4.56 billion years ago.

nonrenewable resource A resource that cannot be replenished or regenerated on the scale of a human lifetime.

normal fault A fault in which the block above the fault surface moves down relative to the block below.

numerical age The age when a rock layer or natural feature was formed, in years before the present.

ocean basins Regions of Earth's crust that are covered by seawater.

ocean conveyor belt *See* **thermohaline circulation**

ocean current A persistent, dominantly horizontal flow of water in the ocean.

oceanic crust The thinner, denser, and younger part of Earth's crust, underlying the ocean basins.

ore deposit A localized concentration in the crust from which one or more minerals can be profitably extracted.

orographic precipitation Precipitation that is induced when moist air is forced over a mountain barrier.

ozone A form of oxygen with a molecule consisting of three atoms of oxygen (O_3).

P wave The first, or primary, wave to be detected by a seismograph.

paleomagnetism The study of rock magnetism in order to determine the intensity and direction of Earth's magnetic field in the past.

paleontology The study of fossils and the record of ancient life on Earth; the use of fossils for the determination of relative ages.

paleoseismology The study of prehistoric earthquakes.

peat A biogenic sediment formed from the accumulation and compaction of plant remains.

percolation The process by which groundwater seeps downward and flows under the influence of gravity.

periglacial Conditions that are near glacial.

permafrost Ground that is perennially below the freezing point of water.

permeability A measure of how easily a solid allows fluids to pass through it.

photosynthesis A chemical reaction whereby plants use light energy to induce carbon dioxide to react with water, producing carbohydrates and oxygen.

plate tectonics The movement and interactions of large fragments of Earth's lithosphere, called *plates*.

pluton Any body of intrusive igneous rock, regardless of size or shape.

plutonic rock An igneous rock formed underground from magma.

porosity The percentage of the total volume of a body of rock or regolith that consists of open space.

precipitation The process by which water that has condensed in the atmosphere falls back to the surface as rain, snow, or hail.

pressure A particular kind of stress in which forces acting on a body are the same in all directions.

pressure gradient A change of atmospheric pressure, measured along a line at right angles to the isobars.

prokaryotes Single-celled organisms with no distinct nucleus—that is, no membrane separates their DNA from the rest of the cell.

proton–proton chain A chain of nuclear reactions in which hydrogen nuclei (or protons) fuse together to form helium nuclei, releasing energy.

pyroclastic flow A stream of hot volcanic fragments (tephra) that are buoyed by heat and volcanic gases and flow very rapidly.

quartzite The product of metamorphism formed by recrystallization of sandstone.

radioactivity A process in which an element spontaneously transforms itself into another isotope of the same element, or into a different element.

radiometric dating The use of naturally occurring radioactive isotopes to determine the numerical ages of minerals, rocks, and fossils.

recharge Replenishment of groundwater.

reef A hard structure on a shallow ocean floor, usually but not always built by coral.

reflection The bouncing back of a wave from an interface between two different materials.

refraction The bending of a wave as it passes from one material into another material, through which it travels at a different speed.

regolith A loose layer of broken rock and mineral fragments that covers most of Earth's surface.

relative age The age of a rock layer, fossil, or other natural feature relative to another feature.

renewable resource A resource that can be replenished or regenerated on the scale of a human lifetime.

reverse fault A fault in which the block on top of the fault moves up and over the block on the bottom.

Richter magnitude scale A scale of earthquake intensity based on the heights, or amplitudes, of the seismic waves recorded on a seismograph.

rock A naturally formed, coherent aggregate of minerals and possibly other nonmineral matter.

rock cycle The set of crustal processes that form new rock, modify it, transport it, and break it down.

Rossby waves Horizontal undulations in the flow path of the upper-air westerlies; also known as *upper-air waves.*

S wave The second kind of body wave to be detected by a seismograph.

salinity A measure of the salt content of a solution.

saltation Sediment transport in which particles move forward in a series of short jumps along arc-shaped paths.

sand A sediment made of relatively coarse mineral grains.

sandstone A medium-grained clastic sedimentary rock in which the clasts are typically, but not necessarily, dominated by quartz grains.

schist A high-grade metamorphic rock with pronounced schistosity, in which individual mineral grains are large enough to be visible.

scientific method The way a scientist approaches a problem; steps include observing, formulating a hypothesis, testing, and evaluating results.

seafloor spreading The processes by which the seafloor splits and moves apart along a midocean ridge and new oceanic crust forms along the ridge.

sedimentary rocks Rocks that form from sediment under conditions of low pressure and low temperature near the surface.

seismic discontinuity A boundary inside Earth where the velocities of seismic waves change abruptly.

seismic wave An elastic shock wave that travels outward in all directions from an earthquake's source.

seismogram The record made by a seismograph.

seismograph An instrument that detects, measures, and records vibrations of Earth's surface.

seismology The scientific study of earthquakes and seismic waves.

shale A very fine-grained fissile or laminated sedimentary rock, consisting primarily of clay-sized particles.

shear A stress that acts in a direction *parallel* to a surface.

shear wave A seismic body wave in which rock is subjected to side-to-side or up-and-down forces, perpendicular to the direction of travel; S wave or secondary wave.

shield volcano A broad volcano with gently sloping sides, built of successive flows of low-viscosity lava, generally of basaltic composition.

slate A very fine-grained, low-grade, metamorphic rock with slaty cleavage; the product of metamorphism of shale.

slope failure The falling, slumping, or sliding of relatively coherent masses of rock.

soil horizon One of a succession of zones or layers within a soil profile, each with distinct physical, chemical, and biologic characteristics.

soil profile The sequence of soil horizons from the surface down to the underlying bedrock.

soil The uppermost layer of regolith, which can support rooted plants.

species A population of genetically and/or morphologically similar individuals that can interbreed and produce fertile offspring.

spring Where the water table intersects the land surface, allowing groundwater to flow out.

strain A change in the shape or volume of a rock in response to stress.

stratigraphy The science of rock layers and the processes by which strata are formed.

stratosphere The layer of atmosphere above the troposphere; here temperature increases slowly with altitude.

stratovolcano A volcano composed of solidified lava flows interlayered with pyroclastic material, generally of andesitic or rhyolitic composition. Such volcanoes have steep sides that curve upward.

streak A thin layer of powdered mineral made by rubbing a specimen on an unglazed fragment of porcelain.

stream A body of water that flows downslope along a clearly defined natural pathway.

stress The force acting on a surface, per unit area, which may be greater in certain directions than others.

subduction zone A boundary along which one plate of lithosphere descends into the mantle beneath another plate.

supernova A stellar explosion marking the death of a massive star.

surf The "broken" turbulent water found between a line of breakers and the shore.

surface creep Sediment transport in which the wind causes particles to roll along the ground.

surface runoff Precipitation that drains over the land or in stream channels.

surface wave A seismic wave that travels along Earth's surface.

suspended load Sediment that is carried in suspension by a flowing stream of water or wind.

suspension Sediment transport in which the wind carries very fine particles over long distances and periods of time.

system A portion of the universe that can be separated from the rest of the universe for the purpose of observing changes that happen in it.

tectonic cycle Movements and interactions of the lithosphere by which rocks are cycled from the mantle to the crust and back; includes earthquakes, volcanism, and plate motion, driven by convection in the mantle.

tension A stress that acts in a direction perpendicular to and *away* from a surface.

theory A hypothesis that has been tested and is strongly supported by experimentation, observation, and scientific evidence.

thermohaline circulation The rising and sinking of water driven by contrasts in water density created by differences in temperature and salinity; this circulation is also known as the *ocean conveyor belt*.

thrust fault A reverse fault that cuts Earth's surface at a shallow angle.

thunderstorms Intense local storms associated with a tall, dense cumulonimbus cloud containing very strong updrafts.

tide A regular, daily cycle of rising and falling sea level that results from the gravitational attraction between the Moon, the Sun, and Earth.

till A heterogeneous mixture of crushed rock, sand, pebbles, cobbles, and boulders deposited by a glacier.

trace fossil Fossilized evidence of an organism's life processes, such as tracks, footprints, and burrows.

trade winds Surface winds that blow from about 30° north and south latitude toward the intertropical convergence zone.

transform fault An approximately vertical fracture in the lithosphere along which two plates slide past each other.

transpiration The process by which water taken up by plants passes directly into the atmosphere.

tropical cyclone An intense traveling cyclone occurring in tropical and subtropical latitudes, accompanied by high winds and heavy rainfall.

troposphere The lowest layer of the atmosphere, in which temperature falls steadily with increasing altitude.

tsunami A train of sea waves traveling over the ocean surface, triggered by an earthquake or other disturbance of the seafloor.

turbidite A graded layer of sediment that is deposited by a turbidity current.

turbidity current A gravity-driven current consisting of a dilute mixture of sediment and water with a density greater than that of the surrounding water.

unconformity A substantial gap in a stratigraphic sequence that marks the absence of part of the rock record.

uniformitarianism The concept that the processes governing the Earth system have operated in a similar manner throughout Earth's history and that past events can be explained by phenomena and forces observable today.

viscosity The degree to which a substance resists flow; a less viscous liquid is runny and flows rapidly, whereas a more viscous liquid is thick and flows slowly.

volcanic rock An igneous rock formed from lava.

volcano A vent through which magma, rock debris, volcanic ash, and gases erupt from Earth's crust to its surface.

water table The top surface of the saturated zone.

wave cyclone Traveling cyclone of the midlatitudes involving interaction of cold and warm air masses along sharply defined fronts.

wave-cut cliff A coastal cliff cut by wave action at the base of a rocky coast.

weathering The chemical and physical breakdown of rock exposed to air, moisture, and living organisms.

Younger Dryas A cold period that occurred between about 11,000 and 10,000 years ago, during the generally mild epoch.

PHOTO CREDITS

Chapter 1

Page 2: Yann Arthus-Bertrand/Photo Researchers, Inc.; (inset) Courtesy NASA; NG Maps; page 3: (top) John S. Shelton; (bottom) James L. Amos/NG Image Collection; page 5: Tom & Susan Bean, Inc./ PO; (left) S.C. Porter; (center) John S. Shelton; (right) Stephen Porter; page 7: (inset) age fototstock/SuperStock, Inc.; Courtesy NASA; page 9: Todd Gipstein/NG Image Collection; page 10: Annie Griffiths Belt/NG Image Collection; page 16: NG Maps; (top left) Marvin Mattelson/NG Image Collection; (center) Dorling Kindersley/Getty Images; (bottom left) Publiphoto/Photo Researchers, Inc.; page 17: (bottom right) Publiphoto/Photo Researchers, Inc.; (top) Publiphoto/Photo Researchers, Inc.; page 18: Frans Lanting/NG Image Collection; page 20: James L. Amos/NG Image Collection; page 21: (left) bildagentur-online.com/th-foto//Alamy; (right) © Nick Brooks; (inset) © Nick Brooks; page 23: (top) Michael Nichols/NG Image Collection; (bottom) James P. Blair/NG Image Collection; page 24: (top) Alberto Garcia/Corbis; (top inset) NG Maps; (bottom) David Rydevik; (bottom inset) NG Maps; page 26: NG Maps; 28: Todd Gipstein/NG Image Collection; page 30: (top) David Rydevik; (bottom) Michael Nichols/NG Image Collection; page 31: Thomas J. Abercrombie/NG Image Collection; page 33: Frans Lanting/NG Image Collection.

Chapter 2

Page 34: Reuters/Landov; (inset) NG Maps; page 35: (top) E.R. Degginger/Photo Researchers, Inc.; (top center) The Natural History Museum; (bottom center) © Breck P. Kent; (bottom) James L. Amos/NG Image Collection; page 37: (top) E.R. Degginger/Photo Researchers, Inc.; (center) Mark A. Schneider/Photo Researchers, Inc.; (bottom) JAMES P. BLAIR/NG Image Collection; page 39: (top left) C.D. Winters/Photo Researchers, Inc.; (top right) Corbis Digital Stock; (top center left) E.R. Degginger/Photo Researchers, Inc.; (top center right) PhotoDisc/Getty Images; (bottom center right) With permission of the Royal Ontario Museum cROM; (bottom center left) David Doubilet/NG Image Collection; (bottom left) Mark A. Schneider/Photo Researchers, Inc.; (bottom right) Corbis Digital Stock; page 40: (top left) MICHAEL S. QUINTON/NG Image Collection; (bottom left) TODD GIPSTEIN/NG Image Collection; (top right) ED GEORGE/NG Image Collection; (bottom right) NG Image Collection; page 41: (top left) JAMES P. BLAIR/NG Image Collection; (bottom left) Digital Vision; (top right) C.D. Winters/Photo Researchers, Inc.; (bottom right) Alice Millikan/iStockphoto; page 42: (top) Courtesy Michael Hochella; (bottom) The Natural History Museum; page 43: (top) Brian J. Skinner; (center) Brian J. Skinner; (bottom) Brian J. Skinner; (right) William Sacco; page 44: (top) William Sacco; (bottom) Brian J. Skinner; page 45: (top) © Breck Kent//Animals Animals; (bottom) William Sacco; page 46: William Sacco; (right) The Natural History Museum; page 47: (center) Aldo Tutino/Art Resource; (bottom left) cAP/Wide World Photos; (bottom right) Xinhua/Landov LLC; page 48: William Sacco; page 51: (top) Brian J. Skinner; (second from top) © Boltin Picture Library; (third from top) © Breck P. Kent; (center) © Breck P. Kent; (third from bottom) © Breck P. Kent; (second from bottom) © Breck P. Kent; (bottom) © Breck P. Kent; page 53: James L. Amos/NG Image Collection; page 55: Kevin Telmer; page 56: (top left) Javier Trueba/MSF/Photo Researchers, Inc.; (top right) Javier Trueba/MSF/Photo Researchers, Inc.; (bottom) Javier Trueba/MSF/Photo Researchers, Inc.; page 57: (top) With permission of the Royal Ontario Museum cROM; (bottom) William Sacco; page 58: (top) Kevin Telmer; (bottom) © Breck P. Kent; page 59: NG Image Collection; page 60: The Natural History Museum.

Chapter 3

Page 62: Richard Nowitz/NG Image Collection; (inset) Gerry Ellis/Minden Pictures/NG Image Collection; NG Maps; page 63: (top) NG Image Collection; (top center) © J.D. Griggs//USGS; (center) Nicole Duplaix/NG Image Collection; (bottom center) © Breck P. Kent; (bottom) © Ken Lucas/Visuals Unlimited; page 64: (top) NG Image Collection; (center) Brian Skinner; (bottom) NG Image Collection; page 65: (bottom left) Brian Skinner; (bottom) NG Image Collection; (center) Brian Skinner; page 67: (left)

Kenneth Garrett/NG Image Collection; (center) Brian J. Skinner; (right) © Tony Waltham; page 68: (right) William Sacco; (left) Brian J. Skinner; page 69: Marc Moritsch/NG Image Collection; page 70: (top) Brian J. Skinner; (center) Brian J. Skinner; (bottom) Brian J. Skinner; page 71: (top left) Carsten Peter/NG Image Collection; (center) © J.D. Griggs//USGS; (top right) Schofield/Getty Images, Inc; page 73: (left) Gerals & Buff Corsi/Visuals Unlimited; (center left) Scientifica/Visuals Unlimited; (right) Ken Lucas/Visuals Unlimited; (center right) Emory Kristof/NG Image Collection; page 75: (left) Jim Richardson/NG Image Collection; (right) Courtesy Dr. Kenneth L. Finger and Chevron Corporation; page 76: Brian J. Skinner; page 77: Nicole Duplaix/NG Image Collection; page 78: (top left) S.C. Porter; (top right) © Stephen Trimble//DRK Photo; (bottom left) Medford Taylor/NG Image Collection; (bottom right) S.C. Porter; page 79: (top left) Minden Pictures, Inc.; (top right) © Tom Bean; (bottom) Courtesy The Field Museum. Photo by Mark Widhalm; page 81: (top) SIPA/NewsCom; (bottom) Lealisa Westerhoff/AFP/Getty Images; page 82: James L. Amos/NG Image Collection; page 83: Courtesy Jay Ague, Dept. of Geology and Geophysics at Yale University; page 85: (top left) William Sacco; (top center) William Sacco; (bottom center) William Sacco; (bottom) William Sacco; page 86: (right) © William Sacco; (left) © William Sacco; page 87: (top left) Brian J. Skinner; (top center) Brian J. Skinner; (top right) Brian J. Skinner; (bottom left) Brian J. Skinner; (bottom right) Brian J. Skinner; page 88: (left) © Breck P. Kent; (left inset) © Runk/ Schoenberger/Grant Heilman Photography; (right) © A.J. Copley/Visuals Unlimited; (right inset) © Craig Johnson; page 89: J.A. Wilkinson//Valan Photos; page 90: (left) © Ken Lucas/Visuals Unlimited; page 91: (top left) Stacy Gold/NG Image Collection; (bottom left) George F. Mobley/NG Image Collection; (top right) Stephen Sharnoff/NG Image Collection; page 92: (top) NG Image Collection; (bottom) Marc Moritsch/NG Image Collection; page 93: © Stephen Trimble//DRK Photo; page 94: (top) © William Sacco; (bottom) J.A. Wilkinson//Valan Photos; page 95: Chris Rainier/NG Image Collection; page 96: (left) Brian J. Skinner; (right) cTony Waltham; (top right) © Lee Boltin/Boltin Picture Library; (center right) © Fred Hirschmann; (bottom right) Richard J. Stewart; page 97: (top) © Breck P. Kent; (top inset) © Runk/ Schoenberger/Grant Heilman Photography; (bottom) © A.J. Copley/Visuals Unlimited; (bottom inset) © Craig Johnson.

Chapter 4

Page 98: Sandy Felsenthal/NG Image Collection; (bottom inset) Associated Press; (top inset) Coinery/Alamy; page 99: (top) Maria Stenzel/NG Image Collection; (top center) Photo Researchers, Inc.; (bottom center) S.C. Porter; (bottom) © William E. Ferguson; page 100: © William E. Ferguson; page 101: Peter Essick/NG Image Collection; page 103: (top right) George F. Mobley/NG Image Collection; (bottom right) Maria Stenzel/NG Image Collection; (top left) Phil Schermeister/NG Image Collection; (bottom right) © Kenneth W. Fink//Ardea London; page 104: (left) Courtesy Brian J. Skinner; (right) Courtesy Brian J. Skinner; page 105: (top left) Brian J. Skinner; (bottom) Thomas J. Abercrombie/NG Image Collection; page 106: (top left) © Gordon Wiltsie/AlpenImages Ltd.; (top right) Stephanie Maze/NG Image Collection; (bottom) Medford Taylor/NG Image Collection; page 107: (top) Courtesy Amy Larson/Smith College; (center) Raymond Gehman/NG Image Collection; (bottom) © Frans Lanting//Minden Pictures, Inc.; page 110: (left) Photo Researchers, Inc.; (right) Photo Researchers, Inc.; page 112: (right) USDA/ Soil Conservation Service; (left) USDA/ Soil Conservation Service; page 116: (top) Raymond Gehman/NG Image Collection; (bottom) Jim Richardson/NG Image Collection; NG Maps; page 118: Michael Nichols/NG Image Collection; page 119: USGS; page 120: (top left) Brian J. Skinner; (bottom) S.C. Porter; page 121: (left) Bill Hatcher/NG Image Collection; (top) Josef Muench; (bottom) Dr. Marli Miller/Visuals Unlimited; page 123: (top left) George Plafker, U.S Geological Survey; (top right) W.E. Garrett R./NG Image Collection; page 124: (bottom left) © William E. Ferguson; (bottom right) Brian J. Skinner; page 125: (top) Paul Johnson/Index Stock; (center left) Walker Howell/NG

Image Collection; (center right) Stephen Sharnoff/NG Image Collection; page 126: (bottom) © Frans Lanting//Minden Pictures, Inc.; page 127: (top) Raymond Gehman/NG Image Collection; (bottom) Michael Nichols/NG Image Collection; page 128: Brian J. Skinner; Associated Press.

Chapter 5

Page 132: James L. Stanfield/NG Image Collection; (inset) Courtesy Brian Skinner; page 133: (top) Emory Kristof/NG Image Collection; (top center) cMarli Bryant Miller; (bottom center) Tyrone Turner/NG Image Collection; (bottom) Walter Meayers Edwards/NG Image Collection; page 136: (top left) Bill Curtsinger/NG Image Collection; (bottom left) Emory Kristof/NG Image Collection; (top right) © Miroslav Krob/AgeFotostock; (bottom right) Bates Littlehales/NG Image Collection; page 137: Annie Griffiths Belt/NG Image Collection; page 139: (top left) George F. Mobley/NG Image Collection; (right) James P. Blair/NG Image Collection; (bottom left) Nicolas Reynard/NG Image Collection; page 141: (bottom left) cMarli Bryant Miller; (right) Visible Earth/NASA; page 142: New York Times Graphics; page 143: (bottom) NASA; page 144: Wes C. Skiles/NG Image Collection; page 145: (left) © Earth Satellite Corporation; (right) © Earth Satellite Corporation; page 147: (inset) Tyrone Turner/NG Image Collection; 147: Tyrone Turner/NG Image Collection; page 150: (left) Peter Essick/NG Image Collection; (bottom left) NGS Maps; page 151: (main) NGS Maps; (bottom) NGS Maps; Ed Kashi/cCorbis; page 152: NG Maps; (top) Phil Schermeister/NG Image Collection; (bottom left) © 2005, Greg Reis; (bottom right) Craig Aurness/CORBIS; NG Maps; page 157: (bottom left) Walter Meayers Edwards/NG Image Collection; page 159: (left) Peter Morgan/Reuters/Corbis; (right) Associated Press; page 161: (left) Wes C. Skiles/NG Image Collection; (right) © Bruno Barbey/Magnum Photos, Inc.; page 162: (top left) Michael Nichols/NG Image Collection; (top right) Michael Nichols/NG Image Collection; (bottom left) Michael Nichols/NG Image Collection; (bottom right) Michael Nichols/NG Image Collection; NG Maps; page 163: Nicolas Reynard/NG Image Collection; page 165: Jim Tuten/Black Star.

Chapter 6

Page 168: Carsten Peter/NG Image Collection; inset John Burcham/NG Image Collection; NG Maps; page 171: GERRY ELLIS/MINDEN PICTURES/NG Image Collection; GORDON WILTSIE/NG Image Collection; page 172: (top left) Carsten Peter/NG Image Collection; (top right) Dr. Cynthia M. Beall & Dr. Melvyn C. Goldstein/NG Image Collection; (bottom left) Marc Moritsch/NG Image Collection; (bottom right) Annie Griffiths Belt/NG Image Collection; page 173: cAP/Wide World Photos; cAP/Wide World Photos; page 174: (top right) S.C. Porter; (bottom) Brian J. Skinner; page 175: (left) © Jim Richardson//Woodfin Camp & Associates; (right) S.C. Porter; page 176: (bottom) © Tom Bean//DRK Photo; page 177: (top left) Gerry Ellis/Minden Pictures; (top right) George Steinmetz/NG Image Collection; (bottom left) George Steinmetz/NG Image Collection; (bottom center) © John S. Shelton; (bottom right) Marc Moritsch/NG Image Collection; page 178: (top left) NASA/JPL/University of Arizona; (top right) NASA/JPL/University of Arizona; 179: Courtesy Stephen C. Porter; page 180: Steve McCurry/NG Image Collection; page 181: NOAA George E. Marsh Album; page 182: Courtesy NASA; page 184: (top) Melissa Farlow/NG Image Collection; (bottom) Frans Lanting/NG Image Collection; page 185: (center) John Lythgoe//Planet Earth Pictures; (bottom right) Peter Essick/NG Image Collection; (top right) Marli Miller/Visuals Unlimited; page 186: Courtesy NASA; page 188: Courtesy Stephen C. Porter; page 190: (center right) Chris Johns/NG Image Collection; (bottom left) Chris Johns/NG Image Collection; page 192: (top) Mark Burnett/Photo Researchers, Inc.; (center) William Thompson/NG Image Collection; (bottom) James P. Blair/NG Image Collection; page 193: (top left) Raymond Gehman/NG Image Collection; (top right) S.C. Porter; (bottom left) Sam Abell/NG Image Collection; (bottom right) CARY WOLINSKY/NG Image Collection; page 194: Maria Stenzel/NG Image Collection; page 195: (top left) Barry Tessman/NG Image Collection; (right) © Marli Bryant Miller; (bottom left) Gordon Wiltsie/NG

Image Collection; NG Maps; page 197: Steve McCutcheon; page 199: (bottom right) CARY WOLINSKY/NG Image Collection.

Chapter 7

Page 200: Galen Rowell/Mountain Light Photography, Inc.; NG Maps; (left) US Navy; (right) US Navy; (top) Emory Kristof/NG Image Collection; page 205: Courtesy Willem van der Westhuizen; page 206: National Library of Wales; page 212: NGS Maps; page 215: (top) Emory Kristof/NG Image Collection; (center left) James L. Stanfield/NG Image Collection; (bottom left) Courtesy NASA; (right) NG Image Collection; page 216: NGS Maps; (bottom) NASA; page 224: DIANE COOK AND LEN JENSHEL/NG Image Collection; (top) Chris Johns/NG Image Collection; (bottom) Hawaii Center for Volcanology; page 227: Emory Kristof/NG Image Collection; page 229: Georg Gerster/Photo Researchers.

Chapter 8

Page 232: Bay Ismoyo/Getty; NG Maps; (top left) DigitalGlobe; (bottom left) DigitalGlobe; page 233: (top) © Lysaght/Liaison Agency, Inc./Getty Images; (center bottom) © Dane Penland/Courtesy Smithsonian Institution; page 235: (left) © John S. Shelton; (right) Winfield Parks/NG Image Collection; page 236: Peter Essick/NG Image Collection; page 239: (top left) © George Plafker, U.S. Geological Survey; (top right) Reza/NG Image Collection; (center right) © Lysaght/Liaison Agency, Inc./Getty Images; (bottom left) Courtesy DigitalGlobe; (bottom right) Courtesy DigitalGlobe; page 241: GRONDIN EMMANUEL/Maxppp/Landov LLC; page 243: cAP/Wide World Photos; page 244: (main) Eric Hanson/NG Image Collection; page 245: (top right) Peter Essick/NG Image Collection; (center right) Courtesy NOAA/NGDC; (bottom right) © Oshihara//Sipa Press; page 246: (center top) Karen Kasmauski/NG Image Collection; (center) Karen Kasmauski/NG Image Collection; (bottom) Waldemar Lindgren/NG Image Collection; page 254: Sarah Leen/NG Image Collection; page 257: (top left) Courtesy Dan Schulze; (bottom left) © Dane Penland/Courtesy Smithsonian Institution; page 258: (right) Bryan & Cherry Alexander/Photo Researchers, Inc.; page 264: Emory Kristof/NG Image Collection; NG Maps; page 267: © Kevin Schafer//Tom Stack & Associates.

Chapter 9

Page 270: Photri/The Image Works; (bottom inset) Chris Newhall//USGS; (top inset) Chris Newhall//USGS; NG Maps; page 271: (top) Carsten Peter/NG Image Collection; (top center) Krafft Explorer/Photo Researchers, Inc.; (bottom center) Brian J. Skinner; (bottom) Raymond Gehman/NG Image Collection; page 272: (left) Paul Chesley/NG Image Collection; (center) Harry Glicken/USGS; (right) Harry Glicken//USGS; page 274: (top left) Frans Lanting/NG Image Collection; (center left) © John S. Shelton; (bottom left) Carsten Peter/NG Image Collection; (top right) © Reuters/Corbis-Bettmann; (bottom right) © Photodisc/SUPERSTOCK; page 275: (bottom) © Steve Vidler/eStock Photo; page 276: (top left) Courtesy Brian Skinner; (left) Photodisc/SUPERSTOCK; (right) Courtesy Brian Skinner; page 277: (left) S.C. Porter; (right) Courtesy Brian Skinner; Paul Chesley/NG Image Collection; page 278: (bottom) © G. Brad Lewis/Liaison Agency, Inc./Getty Images; (top) Roger Rossmeyer/NG Image Collection; page 279: Peter Turnley/Black Star; NG Maps; page 281: Grant Dixon/Lonely Planet Images/Getty; page 282: John Stanmeyer/NG Image Collection; (inset) NG Maps; page 284: NRSC Ltd./Science Photo Library/Photo Researchers, Inc.; page (left) © Science VU-ASIS/Visuals Unlimited; (right) Krafft Explorer/Photo Researchers, Inc.; page 289: (top left) © J.D. Griggs//USGS; (top right) © J.D. Griggs//USGS; page 291: (left) Brian J. Skinner; (right) Brian J. Skinner; page 295: (top) Gary Ladd Photography; (center) Tom Bean/DRK Photo; (bottom) Raymond Gehman/NG Image Collection; page 296: (bottom right) Lyn Topinka/Courtesy USGS; (bottom right) Robert Madden/NG Image Collection; (center right) Jim Richardson/NG Image Collection; (bottom left) P. Frenzen/USDA Forest Service, 1991; NG Maps; (top left) Photodisc/SUPERSTOCK; page 298: (top left) Brian J. Skinner;

(top right) Brian J. Skinner; (bottom) Gary Ladd Photography; page 300: (bottom) © Steve Vidler/eStock Photo; (top) Frans Lanting/NG Image Collection.

Chapter 10

Page 302: © Walter Imber; (inset) Jonathan Blair/NG Image Collection; NG Maps; page 303: (top) NG Image Collection; (center) ROBERT GIUSTI/NG Image Collection; (bottom) JAMES L. AMOS/NG Image Collection; page 305: (top) Annie Griffiths Belt/NG Image Collection; (center) Sam Abell/NG Image Collection; (bottom) CARY WOLINSKY/NG Image Collection; page 306: (top) Walter Meayers Edwards/NG Image Collection; (bottom) © Jeff Gnass; NG Maps; page 307: (left) Landform Slides; (right) Courtesy Lee Gerhard; page 309: Dr. K.Roy Gill; page 310: (top left) NG Image Collection; (top right) NG Image Collection; (bottom) NG Image Collection; page 311: (left) NG Image Collection; (right) Mark Gibson/Index Stock; page 313: Jonathan Blair/NG Image Collection; page 315: (top) MARK HALLETT/NG Image Collection; (center) ROBERT GIUSTI/NG Image Collection; (bottom) KAZUHIKO SANO/NG Image Collection; page 316: (top) KAM MAK/NG Image Collection; (center) KAM MAK/NG Image Collection; (bottom) KAM MAK/NG Image Collection; page 318: (center) American Institute of Physics; (center left) Mary Evans Picture Library/Photo Researchers, Inc.; (bottom right) Mary Evans/Photo Researchers, Inc.; (top) Astrid & Hanns-Frieder Michler/Photo Researchers, Inc.; (bottom) Patrick McFeeley/NG Image Collection; (far right) James P. Blair/NG Image Collection; page 324: (left) Enrico Ferorelli; (right) KENNETH GARRETT/NG Image Collection; page 326: © Sam Bowring; page 327: JAMES L. AMOS/NG Image Collection; page 328: (top) Ralph Lee Hopkins/NG Image Collection; (bottom) Brian J. Skinner; NG Maps; page 329: (top) CARY WOLINSKY/NG Image Collection; (bottom) KAM MAK/NG Image Collection; page 330: JAMES L. AMOS/NG Image Collection; page 331: NG Image Collection; page 332: (center) ROBERT GIUSTI/NG Image Collection.

Chapter 11

Page 334: Photo by psihoyos.com; page 335: (top) O. Louis Mazzatenta/NG Image Collection; (center bottom) Photo by psihoyos.com; (bottom) Chris Johns/NG Image Collection; page 337: (left) © Royalty-Free/CORBIS; page 341: (top left) O. Louis Mazzatenta/NG Image Collection; (center right) Kirk Moldoff/NG Image Collection; (bottom left) © WHOI, D. Foster/Visuals Unlimited; (bottom center) Photo courtesy of Colleen M. Cavanaugh, Department of Organismic and Evolutionary Biology, Harvard University; (bottom right) Woods Hole Oceanographic Institution; page 342: (left) O. Louis Mazzatenta/NG Image Collection; (right) Francois Gohier/Photo Researchers, Inc.; page 343: (left) © Dr. Jeremy Burgess/ SPL/Photo Researchers, Inc.; (right) © CNRI/ SPL/Photo Researchers, Inc.; page 344: (left) O. Louis Mazzatenta/NG Image Collection; (right) O. Louis Mazzatenta/NG Image Collection; page 345: (left) Robert Clark/NG Image Collection; (right) Mary Evans Picture Library/Photo Researchers, Inc.; page 346: (top) NG Maps; (center left) Ralph Lee Hopkins/NG Image Collection; (top right) © J. Dunning/VIREO; (top left) © Gerald & Buff Corsi/Visuals Unlimited; (center right) © Fritz Polking/Visuals Unlimited; (center right) © Joe & Mary Ann McDonald/Visuals Unlimited; (bottom left) Tierbild Okapia/Photo Researchers, Inc.; (bottom right) Eric Hosking/Photo Researchers, Inc.; page 348: (top left) Paul Zahl/NG Image Collection; (top right) Stephen St. John/NG Image Collection; (bottom left) Photo by psihoyos.com; (bottom right) Louie Psihoyos/NG Image Collection; page 349: (left) Courtesy Brian J. Skinner; (right) Robert Clark/NG Image Collection; page 350: Raymond Gehman/NG Image Collection; page 351: (top left) © Edward R. Degginger//Bruce Coleman, Inc.; (top right) Chris Johns/NG Image Collection; (bottom left) © Breck P. Kent/Animals Animals/Earth Scenes; (bottom right) © Theodore Clutter/Photo Researchers, Inc.; page 352: Courtesy Peabody Museum, Yale University; page 353: (left) Hans Fricke/NG Image Collection; (right) Tom McHugh/Photo Researchers, Inc.; page 354: (left) © Francois Gohier/Photo Researchers,

Inc.; (right) Carnegie Museum of Natural History; page 356: (left) © Michael Rothman; (center) Lealisa Westerhoff/Getty Images; (right) © John Reader/ Science Photo Library/Photo Researchers, Inc.; page 358: (top) © Chris Butler; page 359: (top) Jonathan Blair/NG Image Collection; (bottom) Courtesy NASA; page 360: Courtesy Dr. Richard Ernst; page 361: (top) Michael Melford/NG Image Collection; page 363: NGS Maps; page 364: O. Louis Mazzatenta/NG Image Collection; page 365: Paul Zahl/NG Image Collection; page 366: Lealisa Westerhoff/Getty Images; (right) © CNRI/ SPL/Photo Researchers, Inc.; page 367: (top) Jonathan Blair/NG Image Collection; (bottom) Jonathan Blair/Corbis Images.

Chapter 12

Page 370: Pacific Stock/SuperStock; page 372: © Don Dixon; page 375: (top left) Copyright © 1995, David T. Sandwell. Used by permission.; (top right) Copyright © 1995, David T. Sandwell. Used by permission.; page 376: Ken MacDonald/Photo Researchers, Inc.; page 377: (top) Courtesy NASA; (center) Courtesy NASA; (bottom) Courtesy NASA; page 379: Stephen Porter; page 380: (left) Paul Nicklen/NG Image Collection; (right) Tim Laman/NG Image Collection; (bottom) Ingo Arndt/Minden Pictures; page 386: (top left) NG Maps; (center) NG Maps; (bottom) NG Maps; page 387: (top) NG Maps; (bottom right) NG Maps; (bottom left) NG Maps; page 389: (top) Courtesy NOAA; (center) Paul Nicklen/NG Image Collection; (bottom left) David Doubilet/NG Image Collection; (bottom right) Emory Kristof/NG Image Collection; NG Maps; page 390: (top) Copyright © 1995, David T. Sandwell. Used by permission.; page 391: Courtesy Otis B. Brown, Robert Evans, and M. Carle, University of Miami, Rosenstiel School of Marine and Atmospheric Science, Florida, and NOAA/Satellite Data Services Division.

Chapter 13

Page 394: Corbis Images; (inset) Kristoffer Tripplaar/NewsCom; page 395: (top center) Raymond Gehman/NG Image Collection; (bottom center) Tim Laman/NG Image Collection; (bottom) Vie De Lucia/NYT Pictures; page 396: Manley, W.F., 2002, Postglacial Flooding of the Bering Land Bridge: A Geospatial Animation: INSTAAR, University of Colorado, v1, http://instaar.colorado.edu/QGISL/bering_land_bridge; page 397: Manley, W.F., 2002, Postglacial Flooding of the Bering Land Bridge: A Geospatial Animation: INSTAAR, University of Colorado, v1, http://instaar.colorado.edu/QGISL/bering_land_bridge; page 398: (left) © Stephen Rose/Liaison Agency, Inc./Getty Images; (right) Stephen Crowley/New York Times Pictures; page 399: (left) Richard Nowitz/NG Image Collection; This is a MODIS 8-day composite image acquired from 23 October 2008 to 30 October 2008. The image was downloaded from the USGS GLOVIS web-site. This false color image was prepared by Larry Bonneau at the Center for Earth Observation at Yale University.; page 400: (top) Arnulf Husmo/Stone/Getty Images; (bottom) Dr. Susanne Lehner/DLR; page 401: (right) Patrick McFeeley/NG Image Collection; page 402: (top) Raymond Gehman/NG Image Collection; (bottom) © Nicholas DeVore III; page 404: Raymond Gehman/NG Image Collection; page 405: © G. R. Roberts; page 407: (top) David Alan Harvey/NG Image Collection; (center) Cotton Coulson/NG Image Collection; (bottom) Raymond Gehman/NG Image Collection; page 409: Tim Laman/NG Image Collection; page 411: (top) Digital Globe/HO/AFP/Getty Images; (bottom left) Courtesy DigitalGlobe; (bottom right) Courtesy DigitalGlobe; page 412: (left) Richard Reid/NG Image Collection; page 413: (top) Lander, James F. and Patricia A. Lockridge/U.S. Department of Commerce; (bottom) Damian Gadal/iStockphoto; page 414: James P. Blair/NG Image Collection; page 415: (top) James L. Stanfield/NG Image Collection; (bottom) Vie De Lucia/NYT Pictures; page 416: (left) Danita Delimont/Alamy; (right) Danita Delimont/Alamy; page 417: Danita Delimont/Alamy; page 418: (left) Thomas K. Gibson/Florida Keys National Marine Sanctuary; (top right) Phillip Dustan, College of Charleston; (bottom right) © Steven Frink; NG Maps; page 419: (top) Richard Nowitz/NG Image Collection; (bottom left) © Nicholas DeVore III; (bottom right) Raymond Gehman/NG Image Collection; page 420: Digital Globe/HO/AFP/Getty

Images; page 421: Bob Jordan/cAP/Wide World Photos; page 423: Stephen Crowley/New York Times Pictures.

Chapter 14

Page 424: Courtesy NASA; page 425: (top) Todd Gipstein/NG Image Collection; (center) John Dunn/Arctic Light/NG Image Collection; (bottom center) John Eastcott and Yva Momatiuk/NG Image Collection; (bottom) Dick Blume/Syracuse Newspapers//cAP/Wide World Photos; page 426: Todd Gipstein/NG Image Collection; page 427: Peter Hendrie/The Image Bank/Getty Images; page 430: (left) Courtesy NASA; (center) Courtesy NASA; (right) NASA Media Services; page 435: (left) John Dunn/Arctic Light/NG Image Collection; (right) Jeremy Woodhoue/Masterfile; page 440: Adalberto Rias Szalay/Sexto Sol/Photodisc/Getty Images; page 441: (top left) John Eastcott and Yva Momatiuk/NG Image Collection; (top right) Carsten Peter/NG Image Collection; (bottom left) John Eastcott and Yva Momatiuk/NG Image Collection; (bottom right) Todd Gipstein/NG Image Collection; page 442: James P. Blair/NG Image Collection; page 443: Dick Blume/Syracuse Newspapers//cAP/Wide World Photos; page 447: Gandee Vasan/Getty Images; page 448: Warren Faidley/Weatherstock; page 450: (top) Joel Knain/NASA images; (bottom left) NASA image by Jeff Schmaltz, MODIS Rapid Response Team, Goddard Space Flight Center.; (bottom right) Courtesy Terri and Mike Lawson of California, USA/NOAA; NG Maps; page 452: (bottom left) Dick Blume/Syracuse Newspapers//cAP/Wide World Photos; (top right) John Eastcott and Yva Momatiuk/NG Image Collection; page 453: University of Wisconsin Space Science and Engineering Center; page 455: Todd Gipstein/NG Image Collection.

Chapter 15

Page 456: Jaques Descloitres/MODIS Land Rapid Response Team/NASA/Visible earth; NG Maps; page 457: (top) Nick Caloyianis/NG Image Collection; page 458: (bottom) Nick Caloyianis/NG Image Collection; page 461: © Gene Blevins/Los Angeles Daily News/cCorbis; page 470: Courtesy NASA; page 476: Carsten Peter/NG Image Collection; page 478: (bottom left) Joel Sartore/NG Image Collection; page 480: (top) NASA Earth Observatory; page 482: (left) Ping Amranand/SuperStock, Inc.; (right) AP/Wide World Photos; page 485: Steve Pace/Envision.

Chapter 16

Page 488: Science Faction/Getty Images; PHOTOPQR/LA DEPECHE DU MIDI/JEAN LOUIS PRADELS/NewsCom; page 490: NGS Maps; page 491: NGS Maps; page 492: (top right) Courtesy National Academies Press; (center right) Taylor S. Kennedy/NG Image Collection; (bottom right) Courtesy Rob Dunbar, Stanford University; (left) Peter Essick/NG Image Collection; page 493: NASA Images; page 494: Jeffrey Park; page 495: age fotostock/SuperStock, Inc.; page 498: (bottom left) John Dunn/NG Image Collection;

(bottom center) Jim Richardson/NG Image Collection; (bottom right) Tim Laman/NG Image Collection; page 500: (top) William Albert Allard/NG Image Collection; (bottom) © Will & Deni McIntyre/Photo Researchers, Inc.; NG Maps; page 501: (top left) Kari Niemelainen/Alamy Images; page 502: (top) Peter Essick/NG Image Collection; (bottom) Sisse Brimberg/NG Image Collection; page 503: Gerd Ludwig/NG Image Collection; page 504: (top) Hinrich Baesemann/Landov LLC; (bottom) Maria Stenzel/NG Image Collection; page 505: (left) NASA graph by Robert Simmon, based on data provided by the NOAA; (right) NASA graph by Robert Simmon, based on data provided by the NOAA; page 506: (bottom) Ralph Lee Hopkins/NG Image Collection; (top right) Peter Essick/NG Image Collection; (bottom right) cAP/Wide World Photos; page 508: (top right) Peter Essick/NG Image Collection; (top left) Courtesy NASA; (bottom) Stockbyte/SuperStock, Inc.; page 509: (bottom right) cAP/Wide World Photos; (top right) Bertrand Gardel/Hemis/cCorbis; (bottom center) PhotoAlto/SuperStock, Inc.; (top center) James L. Stanfield/NG Image Collection; page 510: (top) Steven Kazlowski/Getty Images; (bottom) Steven Kazlowski/Peter Arnold, Inc.; NG Maps; page 511: (top) Peter Essick/NG Image Collection; (bottom) Jim Richardson/NG Image Collection; page 512: cAP/Wide World Photos; page 513: (bottom) S.C. Porter; page 514: (left) Nasa Images; (right) Jeffrey Park; (bottom) Peter Essick/NG Image Collection.

Chapter 17

Page 516: NASA Images; page 517: (top) © Mary Evans Picture Library; (top center) NASA Images; (bottom) NASA/WMAP Science Team; page 518: © Mary Evans Picture Library; page 520: (top) Bruce Dale/NG Image Collection; page 522: TBGS Observatory; page 523: NASA Images; page 525: NASA Media Services; page 526: C.R. O Dell/Rice University/NASA Media Services; page 528: (top left) Stock Image/SuperStock; (center) NASA/Visible earth; (bottom) NASA/Visible earth; page 529: (top) CONTACT gregm@sierra-remote.com; (center) NASA/Visible earth; (bottom) NASA/Visible earth; page 530: NASA/Visible earth; page 531: NASA/Visible earth; page 532: (left) NASA images; (right) NASA images; 533: (top) NASA images; (bottom) NASA images; page 534: (top left) NASA/JPL; (top right) NASA images; page 535: (left) Breck Kent; (right) Courtesy of the Canada Centre for Remote Sensing Department of Natural Resources Canada; NG Maps; page 537: Tohoku/Getty Images; page 541: (top) Richard Powell; (bottom) ESA AND NASA/NG Image Collection; page 544: (top) cAP/Wide World Photos; (center) NASA/WMAP Science Team; (bottom) NASA / WMAP Science Team; page 545: (left) NASA Images; (center) NASA/JPL-Caltech; (right) NOAO/AURA/Photo Researchers, Inc.; page 546: (top right) NASA Images; (left) NASA/JPL-Caltech/University of Arizon; (right) NASA Images; page 549: NASA Images; page 550: (left) TBGS Observatory; (right) NASA Media Services.

Chapter 1
Figure 1.7: From Flint, Robert F. and Brian J. Skinner, *Physical Geology*, 2nd ed. Copyright 1977 John Wiley & Sons, Inc. Reprinted with permission of John Wiley & Sons, Inc.

Chapter 2
What an Earth Scientist Sees, top: Adapted from Mineral Information Institute, "How Many Minerals and Metals Does It Take to Make a Light Bulb?" Retrieved July 5, 2006, from *www.mii.org/lightbulb.html*. Reprinted by permission of The Mineral Information Institute, *www.mii.org*.

Chapter 5
Figure 5.7: From The New York Times Graphics, "Coastal Defenses Are Disappearing," August 30, 2005. Used by permission of The New York Times Agency. NOTE: Any future revisions, editions thereof in print and any other format require clearance by The New York Times Agency.

Chapter 6
Figure 6.6: Adapted from Pye, K. and L. Tsoar, *Aeolian Sand and Sand Dunes*, Figure 7.1, Chapman & Hall, 1990. Reprinted with kind permission of Springer Science and Business Media; Figure 6.7: Adapted from Hack, John T., *The Geographical Review*, Vol. 31, Figure 19, page 260, by permission of the American Geographical Society.

Chapter 7
Amazing Places: Adapted from Rubin, Ken, "Loihi Volcano—The Bathymetric Map of Loihi Seamount." Hawaii Center for Volcanology web site, 1997. Retrieved October 6, 2006, from *www.soest.hawaii.edu/GG/HCV/loihi.html*. Reprinted by permission of the Hawaii Center for Volcanology, Ken Rubin.

Chapter 8
Case Study: Satake, Kenji, "2004 Sumatra Earthquake," Figure S7, December 2004. Retrieved October 3, 2006, from *http://staff.aist.go.jp/kenji.satake/animation.gif*. Reprinted by permission of K. Satake, Geological Survey of Japan, AIST.

Chapter 9
Figure 9.8: Adapted from Sigurdsson, Haraldur, "Volcanic Pollution and Climate: The 1783 Laki Eruption," *Eos*, Vol. 63, pages 601–602, 1982. Copyright 1982, American Geophysical Union. Modified by permission of The American Geophysical Union.

Chapter 12
Figure 12.1: We have made every effort to secure proper permission for reproducing this image. However, it has not been possible to determine the current copyright owner. If you have information indicating who the copyright owner may be, please contact us at rflahive@wiley.com; Figure 12.11: From Anderson, Bruce and Alan Strahler, *Visualizing Weather and Climate*. Copyright 2009 John Wiley & Sons, Inc. Reprinted with permission of John Wiley & Sons, Inc.; Figure 12.12: National Weather Service.

Chapter 13
Figure 13.11: Drawn by E. Raisz.

Chapter 14
Figure 14.22: From Skaggs, R.H., *Proc. Assoc. American Geographers*, Vol. 6, Figure 2. Used by permission.

Chapter 15
Figure 15.16: Data from U.S. Dept. of Commerce; Figure 15.21: From Anderson, Bruce and Alan Strahler, *Visualizing Weather and Climate*. Copyright 2009 John Wiley & Sons, Inc. Reprinted with permission of John Wiley & Sons, Inc.; Figure 15.23: From Anderson, Bruce and Alan Strahler, *Visualizing Weather and Climate*. Copyright 2009 John Wiley & Sons, Inc. Reprinted with permission of John Wiley & Sons, Inc.

Chapter 16
What an Earth Scientists Sees (B): After Calder, Nigel, *The Weather Machine*, BBC Publications, London 1974. Used by permission of Mr. Nigel Calder.

Chapter 17
Figure 17.22: Adapted from Trefil, James and Robert M. Hazen, *The Sciences: An Integrated Approach*, 5th ed. Copyright 2007 John Wiley & Sons, Inc. Reprinted with permission of John Wiley & Sons, Inc.; Figure 17.22: Adapted from Trefil, James and Robert M. Hazen, *The Sciences: An Integrated Approach*, 5th ed. Copyright 2007 John Wiley & Sons, Inc. Reprinted with permission of John Wiley & Sons, Inc.

Line drawings in the following figures have been adapted from Murck, Barbara, Brian Skinner, and Dana Mackenzie, *Visualizing Geology*. Copyright 2008 John Wiley & Sons, Inc. Reprinted with permission of John Wiley & Sons, Inc.

Chapter 1: 1.4; *What an Earth Scientist Sees*; 1.5; 1.6; 1.8; 1.9; Table 1.1; 1.10; 1.15. **Chapter 2:** 2.1; 2.2; 2.3; 2.5; 2.10; 2.13; 2.14; Table 2.2; 2.15. **Chapter 3:** 3.6; 3.7; 3.13; 3.16; 3.17; *Amazing Places*. **Chapter 4:** *What an Earth Scientist Sees*; 4.4; 4.7; 4.10; 4.11; 4.13; 4.17; *Amazing Places*. **Chapter 5:** 5.1; 5.2; 5.3; 5.8; 5.11; 5.13; 5.15; 5.16; 5.17; 5.18; 5.19; 5.20; 5.21. **Chapter 6:** 6.1; 6.4; 6.5; 6.10; 6.13; 6.15; 6.16; 6.17; 6.19; *What an Earth Scientist Sees*. **Chapter 7:** 7.1; 7.2; 7.3; 7.4; 7.5; 7.6; 7.7; 7.9; *What an Earth Scientist Sees*; 7.11; 7.12; 7.13; 7.15. **Chapter 8:** 8.1; 8.3; 8.4; 8.6; 8.9; 8.10; 8.11; 8.12; 8.13; 8.15; *What an Earth Scientist Sees*; 8.16; 8.17; 8.18; *Amazing Places*. **Chapter 9:** 9.3; *Case Study*; 9.9; 9.13; 9.14; 9.15; 9.17; 9.20; 9.21. **Chapter 10:** 10.3; 10.7; 10.9; 10.11; 10.12; 10.13; 10.14; 10.15; 10.16; *Case Study*. **Chapter 11:** 11.1; 11.2; 11.3; 11.11; 11.19; 11.20; 11.22; *Amazing Places*. **Chapter 12:** 12.9; 12.10; 12.14; 12.15. **Chapter 13:** 13.3; 13.5; *What an Earth Scientist Sees*. **Chapter 14:** 14.1; 14.2; 14.3. **Chapter 15:** 15.13. **Chapter 16:** 16.1; 16.18. **Chapter 17:** 17.9; 17.17.

Line drawings in the following figures have been adapted from Strahler, Alan and Zeeya Merali, *Visualizing Physical Geography*. Copyright 2008 John Wiley & Sons, Inc. Reprinted with permission of John Wiley & Sons, Inc.

Chapter 3: 3.2. **Chapter 4:** 4.8. **Chapter 5:** 5.6; 5.14; 5.23. **Chapter 12:** *What an Earth Scientist Sees*; 12.13. **Chapter 13:** 13.7. **Chapter 14:** *What an Earth Scientist Sees*; 14.6; 14.7; 14.8; 14.10; 14.11; 14.12; 14.13; 14.15; 14.17; 14.19; 14.20; 14.21. **Chapter 15:** 15.2; 15.4; 15.5; 15.7; 15.8; 15.9; 15.10; 15.11; 15.12; 15.14; 15.15; 15.17; 15.18; 15.19; 15.25. **Chapter 16:** 16.7. **Chapter 17:** 17.4.

Line drawings in the following figures have been adapted from Skinner, Brian, Stephen Porter, and Daniel Botkin, *The Blue Planet: An Introduction to Earth System Science*. Copyright 1999 John Wiley & Sons, Inc. Reprinted with permission of John Wiley & Sons, Inc.

Chapter 1: 1.12. **Chapter 4:** 4.12. **Chapter 12:** 12.2; 12.8. **Chapter 14:** 14.5. **Chapter 15:** 15.3. **Chapter 17:** 17.2; 17.5; 17.6; 17.8.

Line drawings in the following figures have been adapted from Skinner, Brian, Stephen Porter, and Jeffrey Park, *Dynamic Earth: An Introduction to Physical Geology*. Copyright 2004 John Wiley & Sons, Inc. Reprinted with permission of John Wiley & Sons, Inc.

Chapter 7: 7.14; 7.16. **Chapter 8:** 8.12. **Chapter 9:** 9.5; 9.19. **Chapter 12:** 12.5; 12.6. **Chapter 16:** 16.4; 16.5.

INDEX